TELEPEN
D0477662

MANAGING THE SYSTEM LIFE CYCLE

YOURDON PRESS COMPUTING SERIES
Ed Yourdon, *Advisor*

BENTON AND WEEKES Program It Right: A Structured Method in BASIC
BLOCK The Politics of Projects
BODDIE Crunch Mode: Building Effective Systems on a Tight Schedule
BRILL Building Controls Into Structured Systems
BRILL Techniques of EDP Project Management: A Book of Readings
CONSTANTINE AND YOURDON Structured Design: Fundamentals of a Discipline of Computer Program and Systems Design
DE MARCO Concise Notes on Software Engineering
DE MARCO Controlling Software Projects: Management, Measurement, and Estimates
DE MARCO Structured Analysis and System Specification
DICKINSON Developing Structured Systems: A Methodology Using Structured Techniques
FLAVIN Fundamental Concepts in Information Modeling
HANSEN Up and Running: A Case Study of Successful Systems Development
KELLER Expert Systems Technology: Development and Application
KELLER The Practice of Structured Analysis: Exploding Myths
KING Creating Effective Software: Computer Program Design Using the Jackson Method
KING Current Practices in Software Development: A Guide to Successful Systems
MAC DONALD Intuition to Implementation: Communicating About Systems Toward a Language of Structure in Data Processing System Development
MC MENAMIN AND PALMER Essential Systems Analysis
ORR Structured Systems Development
PAGE-JONES The Practical Guide to Structured Systems Design
PETERS Software Design: Methods and Techniques
ROESKE The Data Factory: Data Center Operations and Systems Development
SEMPREVIO Teams in Information Systems Development
THOMSETT People and Project Management
WARD Systems Development Without Pain: A User's Guide to Modeling Organizational Patterns
WARD AND MELLOR Structured Development for Real-Time Systems, Volumes I, II, and III
WEAVER Using the Structured Techniques: A Case Study
WEINBERG Structured Analysis
WELLS A Structured Approach to Building Programs: BASIC
WELLS A Structured Approach to Building Programs: COBOL
WELLS A Structured Approach to Building Programs: Pascal
YOURDON Classics in Software Engineering
YOURDON Design of On-Line Computer Systems
YOURDON, LISTER, GANE, AND SARSON Learning to Program in Structured Cobol, Parts 1 and 2
YOURDON Managing Structured Techniques, 3/E
YOURDON Managing the System Life Cycle, 2/E
YOURDON Structured Walkthroughs, 2/E
YOURDON Techniques of Program Structure and Design
YOURDON Writing of the Revolution: Selected Readings on Software Engineering
ZAHN C Notes: A Guide to the C Programming

MANAGING THE SYSTEM LIFE CYCLE

Second Edition

Edward Yourdon

YOURDON PRESS
A Prentice Hall Company
Englewood Cliffs, New Jersey 07632

Library of Congress Cataloging-in-Publication Data

Yourdon, Edward.
Managing the system life cycle.

(Yourdon Press computing series)
Bibliography: p.
Includes index.
1. Computer software—Development—Management.
2. Electronic data processing—Structured techniques.
I. Title. II. Series.
QA76.76.D47 1988 005.1 87-23001
ISBN 0-13-547530-9

Editorial/production supervision
and interior design: Richard Woods and Barbara Marttine
Cover design: Wanda Lubelska Designs
Manufacturing buyer: Richard Washburn

A Division of Simon & Schuster
Englewood Cliffs, New Jersey 07632

The publisher offers discounts on this book when ordered in bulk quantities. For more information, write:

Special Sales/College Marketing
Prentice Hall
College Technical and Reference Division
Englewood Cliffs, NJ 07632

Printed in the United States of America

10 9 8 7 6 5 4 3 2 1

ISBN 0-13-547530-9 025

Prentice-Hall International (UK) Limited, *London*
Prentice-Hall of Australia Pty. Limited, *Sydney*
Prentice-Hall Canada Inc., *Toronto*
Prentice-Hall Hispanoamericana, S.A., *Mexico*
Prentice-Hall of India Private Limited, *New Delhi*
Prentice-Hall of Japan, Inc., *Tokyo*
Simon & Schuster Asia Pte. Ltd., *Singapore*
Editora Prentice-Hall do Brasil, Ltda., *Rio de Janeiro*

Contents

Preface

Much has happened since the first edition of this book was published in 1982. This is inevitable in the computer field, of course, and I expect that significant changes will continue to occur through the rest of this decade and well into the 1990s.

This book is about the management of systems development projects that use such "structured" techniques as structured analysis, structured design, and structured programming. When I began writing the first edition of this book around 1980, structured programming was widely accepted and structured design was widely discussed, but structured analysis—probably the most important of the structured techniques—was not in common use at all. Now, in the late 1980s, all these structured techniques are well understood by computer programmers and systems analysts and have been used to help build thousands of computer systems around the world.

However, the structured techniques themselves have evolved and changed during this period, and those changes are reflected in this new edition. Probably the most important change is the incorporation of *data* modeling into a set of systems analysis techniques that previously concentrated almost exclusively on *function* modeling; today's system models include, along with the familiar data flow diagrams, various forms of *entity relationship diagrams,* which replace the data structured diagrams associated with the "classical" structured techniques of the late 1970s and early 1980s. Modern structured analysis also has tools for modeling the time-dependent behavior of systems (using *state transition diagrams,* sometimes known as "finite-state diagrams") and can thus be used on real-time systems as well as on the classical business-oriented systems.

There have also been changes in the "methodology" of structured analysis. The first edition of this book emphasized a series of analysis activities that led to four distinct models of a system: the "current physical," the "current logical," the "new logical," and the "new physical." Although there were good reasons for asking the analyst to build these models, our experience in the real-world use of these methods in hundreds of consulting projects during the past several years has shown us the danger of spending too much time modeling the user's current system—a system that will of necessity be thrown away and replaced. Consequently, this edition reflects the current view of structured analysis, in which the analyst is asked to begin with a model of what the user's *new* system must do.

Similarly, "classical" structured techniques of the late 1970s and early 1980s suggested that the model of user requirements could be transformed directly into a model of code organizations; this was because most systems built in those ancient times tended to be implemented as monolithic batch programs running on a single mainframe computer. This edition takes note of the fact that many systems are implemented on a network of computers and on a network of communicating processes (or "tasks" or "partitions," etc.) within each processor. Hence there are two intermediate design modeling activities (one at the processor level, and one at the task level) before the conventional structured design activity of producing structure charts is reached.

The concept of *prototyping* has also become popular in the past few years and is discussed in this book. Though prototyping tools are an important addition to the "tool kit" available to the project team, I continue to be concerned that prototyping is being sold (often by the vendors who created the software tools) as a panacea that will replace much of the work of structured analysis with instant gratification. I remain convinced that many of the benefits of prototyping can be achieved with the structured techniques; this is discussed in Chapter 3.

Finally, we have seen the advent of many *automated tools* for systems analysts and designers in the years since the first edition of this book was published. These tools, developed by such companies as Intech, Nastec, Yourdon, Cadre, and Tektronix, enable the systems analyst and systems designer to automate much of the graphics and text associated with the requirements models and implementation models of a system. One could argue that these tools do not change the project activities themselves or the way that the project activities are managed; indeed, the project team must carry out systems analysis whether or not there is automated support, and the project manager must ensure that the proper "products" of systems analysis are produced at the proper time. However, I am convinced that the *quality* of work produced by the systems analyst and designer will improve dramatically with the use of such tools, and this will (indirectly) improve the manageability of the project.

I have had the honor and good fortune to work with a number of excellent, dedicated data processing professionals during the years since the first edition of this book was published; many of them have offered insights and suggestions that have helped me tremendously in the preparation of this new edition. I am especially grateful to Bob Spurgeon, William Wenker, and Eric Blaustein for their careful reading of the draft manuscript and for their many helpful suggestions. Finally, I would like to thank my wife and children for their patience and forbearance while I sat hunched over my Macintosh computer for days on end; they deserve a great deal of credit for helping create this book.

Introduction

WHAT THIS BOOK IS ABOUT

The purpose of this book is to present a system life cycle for systems development projects that makes use of structured systems development techniques, in sufficient detail for you to use in a real project. In other words, this book is designed to give you the necessary guidelines with which to organize, manage, and control a systems development project—in particular one that makes use of structured analysis, structured design, and structured programming.

Why should you read this book? Presumably because you're looking for a way to organize your systems development projects or because you're not satisfied with the methodology that you currently use. By the end of this book, you will have learned the important *activities* and *products* associated with a systems development project, and you should be able to apply the concepts to your next project.

I also hope to shed some light on an issue that often generates loud, emotional debates in many organizations, the issue of "methodologies." Three questions arise: Should our organization use a methodology? If so, what should it look like? What should its components be?

A *methodology,* as we will use the term in this book, is a formal specification of a *system for building systems.* It defines the pieces, or components, of the system for building computerized information systems—that is, the phases or activities that one finds in a typical software development project. It also defines the interfaces between those components.

You'll also find that this book makes a distinction between a *technical* model of the system for building systems and a *managerial* model. Most methodology products and textbooks on project management concentrate entirely on the latter: They tell the manager how to control and supervise the activities in his project. But as my colleague Tom DeMarco points out, "You can't have a methodology without methods." You can't hope to organize, manage, and control a systems development project unless you can describe to your technicians—systems analysts, systems designers, database designers, and programmers—what you expect them to be doing during various phases of the project and what kind of technical products you expect them to deliver. This book concentrates on that aspect of the methodology and places considerably less emphasis on the classical control and supervisory aspects of project management.

Throughout this book, I assume that you have had some prior exposure to systems development projects, either as a programmer-analyst or as a project leader. I hope that this book will be useful as well to high-level managers, training coordinators, university professors, presidents, and kings—but in my mind's eye, it's the project leader responsible for project management whom I'm addressing as I write these words.

WHAT THIS BOOK IS NOT ALL ABOUT

There are several things that I am definitely *not* attempting to accomplish in this book.

- *This book will not teach you how to be a manager.* If you don't know how to manage and motivate human beings, don't expect any help from this book. If you don't have a basic sense of delegation, organization, and administration, I'm not sure you can obtain help from *any* book; but you certainly won't get any pearls of wisdom from this one! This is, of course, a major issue for the programmer or systems analyst who has just been promoted to the rank of project manager and is faced with his first project. If you are in this situation, consult standard introductory texts on project management before you read this book.
- *This book will not tell you how to organize your EDP or MIS department.* This book is concerned with the activities of an individual systems development project; as you'll see in Chapter 3 and the subsequent chapters, we will talk about the interactions between the project team and management end users (otherwise known as "clients," "customers," or "owners"), and the operations department. But there is little or nothing said about such things as a "steering committee" of users and managers who help set priorities among the various projects in the

organization. Nor is anything said about how the technicians in the EDP or MIS department should be organized for maximum efficiency—whether systems analysts and programmers should be separate groups or whether development teams should be kept separate from maintenance teams. However, the book will provide some comments on the impact of the structured systems development techniques on the typical "life cycle" that an EDP or MIS organization is likely to use.

- *This book will not make you an expert in using structured systems development techniques.* If you've never heard of structured programming or structured analysis, this is not the book to read. There are lots of good books that will give you the basics; the discussion in Chapter 2, for example, provides an overview and some references for additional reading. But for the most part, this book assumes that you're already familiar with the basic concepts of structured analysis, structured design, structured programming, top-down development, and walkthroughs.
- *This book will not make you an expert estimator.* I can't estimate how long it will take to build an underground house in a swamp—yet that's the sort of thing many of our end users and managers really want us to do! Chapter 10 discusses some of the problems of scheduling and estimating, but I should warn you in advance: *There is no magic.* For more information on this area, I recommend that you consult some of the classic books on software metrics, such as those by DeMarco (1982), Boehm (1982), and Jones (1986) in the bibliography.
- *This book will not help you solve political problems.* Virtually everything in this book assumes that you are a rational manager, that you have rational programmers and systems analysts working for you, that you deal with rational users and customers, and that the environment in which you determine your schedules and manpower estimates is not only rational but also friendly and supportive. I assume that you, your subordinates, your users, and your superiors are interested in producing a "quality" information system—one that does what the user wants, and does it in a reasonably economical, reliable, maintainable fashion. If your environment is just the opposite—if, for example, you are rewarded *solely* for finishing your project on time, with no concern whatsoever about the quality of the product—this book probably won't help you. If you have arbitrary and unreasonable deadlines imposed on you from on high, or if you have users who refuse, on principle, to discuss their requirements with you, you're in deep trouble. Rather than reading this book, I suggest that you consider polishing your résumé and looking for a better place to work. As an alternative, consult such books at Block's *Politics of Projects* (1983) or Page-Jones's *Practical Project Management* (1985) for advice and guidance.
- *This book will not provide you with a "religious" view of software*

development. Many textbooks on project management, and most software development methodology packages, take a hard-line approach to the various activities that they identify in the project life cycle—that is, they emphasize that the programmer-analyst *must* carry out tasks A, B, and C in a certain sequence in *all* cases, under *all* conditions. That's understandable in the case of methodology products, whether purchased for $50,000 or developed within the corporation for a cost sometimes approaching $500,000. When that kind of money is involved, the person who commits to the expenditure usually defends the package the same way a wild animal defends its offspring against predators: "If you spend all that money on the XYZ methodology package," says the corporate EDP manager, "you can be damned sure that every single one of my programmers will memorize every page of it and follow it to the letter!" Considering the relatively low cost of this book, there should be no need for religious mania—even if you discard 90 percent of the ideas in the book, it won't have cost you $500,000. Indeed, I will deliberately suggest that you improvise in certain areas and that you fill in the details in other areas based on your own experience, and I hope that you feel equally free to ignore sections of this book that don't apply to your own needs.

THE ORGANIZATION OF THIS BOOK

Now that I've given you a thumbnail sketch of what the book will and will not provide, let me give you a slightly more detailed picture of the individual chapters.

Chapters 1 and 2 serve as an introduction to the main subject. Chapter 1 compares the advantages and disadvantages of conventional systems development projects, semistructured projects, and modern structured projects. Chapter 2 contains a brief overview of the structured techniques themselves, with suggested references in case you need more detail. If you're already familiar with the structured techniques, and if you're feeling eager to proceed, turn directly to Chapter 3.

The heart of the book consists of seven chapters, 3 through 9. Chapter 3 gives an overview of the structured project life cycle and its component activities. Chapters 4 through 8 discuss the major activities: survey, analysis, design, and implementation. Chapter 9 addresses the final major activities, including generation of user manuals, conversion of the database, installation of the system, and quality assurance.

The concluding chapter sums it all up, placing emphasis on the manager's special role and obligations.

1

Changes in Project Management

1.1 INTRODUCTION

As I indicated in the introductory chapter, this book is concerned with a relatively new kind of life cycle for systems development projects—the *structured project life cycle.* It is also concerned with the proper use of structured analysis, structured design, and structured programming in a systems development project. Consequently, there is a strong implication that the new, structured kind of life cycle is in some way better than the conventional project life cycle. And, similarly, that the use of structured analysis and related techniques leads to far more successful projects than does the use of conventional analysis and development techniques. Indeed, one of the premises of this book is that the conventional projects tend to be over budget, behind schedule, expensive to develop, expensive to maintain, unreliable, and unacceptable to users.

Of course, these problems do not apply to *all* conventional EDP projects, but the larger the scope and size of the project, the more likely it is to show evidence of some or all the difficulties just listed. I think it is reasonable to make the following observations about the relationship between the size of a project and its complexity:

SIZE OF PROJECT	LEVEL OF COMPLEXITY
Up to 1,000 lines	Trivial
1,000 to 10,000 lines	Simple
10,001 to 100,000 lines	Difficult
100,001 to 1,000,000 lines	Complex
1,000,001 to 10,000,000 lines	Nearly impossible
More than 10,000,000 lines	Utterly absurd

(Obviously, lines of code is only one measure of complexity, and a primitive one at best. However, it is adequate to illustrate my point in this chapter.)

Trivial projects, defined in terms of lines of code, may not require any formal project management. A computer program or system that involves only a few hundred lines of code can usually be implemented by one person in a period of a few days to a few weeks. Generally, all one wants to know at the beginning of such projects is the deadline: "When is it going to be finished?" And even though structured analysis, structured design, and structured programming can be of great benefit even in such small projects, the programmer—by brute force or just common sense—can get the job done using conventional methods. If problems arise and the schedule slips, the manager can always fall back on what my colleague Tom DeMarco calls "the time-honored tradition of unpaid overtime": The manager can gently persuade the programmer to work nights and weekends until the project is finished.

Indeed, much the same thing can be said about *simple* projects, those involving up to 10,000 lines of code. Such a project typically involves three or four programmer-analysts for a period of 6 to 12 months. It's large enough, and it lasts long enough, for some kind of formal project management to be necessary, but it's the sort of project that a veteran manager has experienced dozens of times in his career. Precisely because it *is* within the manager's realm of comprehension, he is able to organize, manage, and control it. And although the project life cycle specified in this book is quite useful, managers will argue that implementing such a formal discipline represents unnecessarily heavy artillery. Similarly, although the techniques of structured analysis, structured design, and structured programming will lead to a more maintainable product, their use may not speed up development time at all.[1]

However, projects involving between 10,001 and 100,000 lines of code—projects that I categorize as *difficult*—are almost beyond the ability of the manager to handle easily. Such a project can involve half a dozen to a dozen programmers and can last two or three years. Formal project organization

[1]In general, proper use of the structured techniques will improve the productivity of the development phase of a system by approximately 10 to 15 percent. However, maintenance costs are usually reduced by a very substantial amount, usually ranging from a factor of 2 to a factor of 10.

is an obvious necessity, and a formal approach to the analysis and design of the EDP system is also usually essential. In most medium to large EDP organizations, a project of this size will eventually finish—if only because the organization continues to pour people and money into the project until it staggers through to completion.[2] However, there is a nontrivial danger that the project will be finished several months late and that it will exceed its budget by tens of thousands or even hundreds of thousands of dollars. Consequently, the techniques of structured analysis, design, and programming become critically important; with the proper use of these techniques, there is a reasonable chance that the project can be completed on time and within the stated budget.

Perhaps more important than use of the techniques is the use of the *structured project life cycle* in projects of this size. A project involving up to 100,000 lines of code is sufficiently complex that neither the systems analyst nor the user is likely to have a crystal-clear understanding of exactly what the system is supposed to do, and even if there *is* a clear understanding, the user is likely to change his mind about some of the requirements. Also, if the project lasts two or three years, there is a reasonable chance that the environment—that is, the technology, the local business conditions, the applicable government regulations, and even the user himself—will change by the time the project is finished. Nevertheless, a veteran project manager may tell you that none of these problems is insurmountable, that it's all a question of "good management."

With *complex* projects, involving between 100,001 and 1,000,000 lines of code, even the most battle-scarred veteran would admit to a certain amount of nervousness. At this level of complexity, 50 to 100 people may be involved in a project that can last three to five years or more. On such a project, many members of the project team either will leave before it is finished or will be hired in the middle, and there will be at least two levels of project management. In addition, we can be reasonably sure that the system is not being built for just one user or even one homogeneous group of users. Instead, there are likely to be diverse (and often conflicting) communities of users, each of which has its own local view of the requirements of the system.

If conventional development techniques are used in a project life cycle, there is a good chance that (a) the project will never be finished, (b) it will be finished so far behind schedule and so much over budget that it will damage or ruin the manager's career, (c) the user will reject the final system as being unsuited to his needs, or (d) all of the above. Although the techniques espoused

[2]However, in today's turbulent business world, the user may not be willing to continue spending money on a project once it gets substantially behind schedule or over budget. Various surveys, including one by T. Capers Jones (1986), have reported that as many as 25 percent of all DP projects never finish!

in this book are not foolproof—that is, they do not guarantee success if the project is under the control of an incompetent manager and inexperienced programmers and analysts—they *are* essential if the manager is to finish the project on time and within budget.

The next category, *nearly impossible,* is one that few managers have experienced. Only a handful of projects have ever involved between 1 million and 10 million lines of code.[3] Conventional techniques and a conventional project life cycle are almost guaranteed to fail with projects of this magnitude; indeed, even with the techniques discussed in this book, the normal problems of politics, personnel manager, and the vagaries of fate will make successful project completion nearly impossible.

A current example of the nearly impossible project is the so-called Star Wars defense program, which is estimated to require several tens of millions of lines of code operating in a complex multicomputer, real-time environment. Many of the world's leading computer scientists have expressed grave doubts that such a system can *ever* be successfully implemented; indeed, many would argue that the Star Wars project would be more accurately characterized as "utterly absurd."

1.2 CHARACTERISTICS OF CONVENTIONAL EDP PROJECTS

Why is it that the conventional approach to managing and developing EDP systems has produced such a dismal record of failure on all but trivial projects? Among the basic causes for failure are these three:

- Poor systems analysis
- Little or no control over design and code
- Bottom-up development and integration

In the paragraphs that follow, I shall explain my reasons for these assertions.

First, let's consider poor systems analysis. Perhaps the most serious problem has been in this area because, after all, even the most brilliant design and the most elegant code are of no use if the requirements of the system have not been stated well. And the larger the project, the more difficult it is to state

[3]That is, there have only been a handful of projects that were estimated *when they began* to require 1 million to 10 million lines of code. Some large organizations, such as banks and insurance companies, have found that during the past 10 to 20 years they have accumulated a total of several million lines of code, and their users now insisting that *all* of the code function together as an integrated, on-line, real-time, general-purpose management information system. Such cases are substantially different from a project that at the beginning is expected to involve up to 10 million lines of code.

the requirements of the system in such a way that they can be clearly understood by user and analyst alike. In most conventional projects, systems analysts produce a voluminous tome of narrative English,[4] which the users don't actually read and which they probably wouldn't understand even if they did. Compounding the trouble is the difficulty inherent in the task of modifying or updating the huge specification. Pity the poor user who wants to make a change to the requirements after the specifications have been frozen!

Second, many conventional projects have little or no control over the quality of the design or the code. I have had many veteran technicians and managers explain to me, "If it works, that's good enough. We don't have time to dream up seventeen different designs and figure out which one is best!" The practical consequence of this attitude is that each programmer and systems analyst tends to develop a personal style, with no objective way of comparing the quality of work being performed by each person. And with no objective way of determining quality, there is also no realistic way of measuring progress and determining how much of the technician's job is finished—after all, anyone can finish a job within an allotted time by doing the job shoddily.

A third problem occurs because conventional systems are often built from the bottom up. That is, after the system has been designed and coded, integration proceeds by carrying out *module* testing first, followed by *program* testing and then by various levels of *subsystem* testing, all of which is concluded by *system* testing. We will discuss the problems with this approach in more detail in Chapter 3.

All of the problems listed here are *technical* problems that can be remedied by the proper use of structured analysis, structured design, and structured programming. From a *management* point of view, however, the disadvantages are somewhat different and occur in conjunction with the following aspects of projects:

- Sequential or one-shot project phases, which provide no opportunity for iteration
- Overemphasis on plans and checklists, giving the manager the impression that he can avoid difficult decision-making activities
- Unavailability of evaluation criteria, needed to determine whether critical activities are being performed properly
- Impracticality of completeness proofs for determining whether the activities were ever done

I have listed the first of the management problems as resulting from sequential project phases. By this I simply mean that most conventional projects insist that while, say, the phase of *analysis* is taking place, no other activity

[4]Or French or German or Russian or Chinese or Swahili or...

can take place. This means that design, the next phase in the development cycle, should not commence until analysis is *completely* finished. Such an approach can be legislated, but it can never be totally enforced: The project team will inevitably engage in some design work (even if it is "unofficial") before the analysis phase of the project is finished. In any case, such a sequential approach, by its nature, precludes the opportunity to "overlap" activities and thus reduce the elapsed calendar time of the project. An even less tenable situation occurs in conventional projects that hold that once the analysis phase is finished, no more analysis is supposed to take place for the remainder of the project.

This approach may have a certain appeal on small or medium-sized projects because the manager knows that only one thing is going on at any given time and that when the activity is finished, it is truly finished. On large projects, such an approach defies reality. It is impractical (and usually impossible) to carry out an activity like analysis without *simultaneously* considering some of the consequences of the analysis decisions on the design of the system. And it's foolish to insist that *all* of the activities of analysis be finished before a number of the subsequent activities of design, coding, and testing be allowed to begin. After all, the analysis phase alone could require years of work, and the user probably won't tolerate that for long without seeing that something is being accomplished. The worst sin of all is assuming that once an activity is done, it's done for good. Few of us are rash enough to boast that we can do an exceedingly complex job right the first time, so why should we allow our project life cycles to insist on this one-shot approach?

Overemphasis on plans and checklists is a problem with conventional project management. I don't deny the need for plans and checklists; it's just that I don't think they are a panacea. Several consulting firms have grown quite rich during the past decade by selling 1000-page project management manuals—sometimes for $100,000 or more—that in my view do little more than give the project manager a false sense of confidence. Few managers can really believe that the programmers and analysts read all 1000 pages of detailed instructions on how to conduct a project. Let's face it, even a sex manual would be boring if it were 1000 pages long! In addition, these project management products often require substantial time and effort by the programmers and analysts in order to fill out the required forms; this is time that might otherwise be spent doing useful work, and thus project productivity is often reduced substantially. And since members of the technical staff view the paperwork as uninteresting and unproductive, they resent the effort required—thus leading to the provision of inaccurate or downright false information.[5]

[5]This is a variation on the principle known as the "Heisenberg uncertainty principle" in nuclear physics: Measuring a phenomenon (in the case of nuclear physics, either the mass or the velocity of an atomic particle) changes the phenomenon. In the case of software project management, measuring the amount of time spent carrying out various activities is itself a time-consuming activity. The only solution available now is workstation-based development tools (programmer workbenches) where the technician's activities are automatically (and invisibly) monitored and measured by the workstation.

The third failure of conventional projects is the lack of objective criteria with which to judge whether the critical activities of analysis, design, and coding are being carried out properly. In the worst case, this deficiency means that every programmer and every analyst can perform the job according to personal whim; even if you, as a manager, try to impose quality control, you're likely to hear emotional arguments from your technicians—arguments in which they try to defend their years of experience or their common sense against the technical criticisms of others. Without techniques like those of structured analysis, structured design, and structured programming (which, if the terms are used properly, mean the same thing to all technicians), there is a danger that you and your staff will build an idiosyncratic product, difficult for anyone else to maintain or modify. Even worse, the various pieces of the system will reflect the idiosyncrasies of their authors and consequently may not fit together.

1.3 CHARACTERISTICS OF STRUCTURED PROJECTS

As you may have guessed, the characteristics of structured projects are essentially opposite to those of conventional projects. Specifically, structured projects are characterized by better tools for expressing user rquirements, emphasis on quality design and good code, and top-down systems development. All of these benefits come from the use of structured analysis, design, and programming techniques, which we will discuss in more detail in Chapter 2.

From a management point of view, however, we can describe structured projects as having the following features:

- Meaningful paper models of the system to be built
- Visibility of the analysis, design, and programming effort
- Objective criteria for measuring and determining "goodness" or quality
- An iterative approach to the activities of building an EDP system

The concept of a paper model is central to the structured techniques and in addition has important management ramifications. In most engineering fields it is common to develop a *model* of the system to be built, something that both users and managers can use to gauge the viability of the system. Used in conjunction with tools like data flow diagrams, entity relationship diagrams, state transition diagrams, data dictionaries, structure charts, and structured English, modeling has finally become practicable in the data processing field, too.

A major advantage of a paper model is that it makes the process of analysis, design, and coding more visible to the manager. Instead of being a mysterious activity that takes place primarily within the technician's cranium, it's out in the open so the manager can see whether work is proceeding on schedule; among other things, this helps avoid the "99 percent complete" syndrome seen in many systems development projects.

Another advantage of the structured approach is that it contains objective criteria for determining what is "good" analysis, design, and coding. Perhaps most important is that the techniques encourage *uniformity* and *consistency* in the analysis, design, and coding activities, rather than allowing every technician to define a personal, idiosyncratic version of "good." This is particularly important for the project that uses walkthroughs as a form of peer group reviews of a product.

Finally, the structured approach emphasizes a top-down iterative approach to the analysis, design, and implementation of large, complex EDP systems. It assumes that some "early" activities, such as systems analysis, will be incomplete when other, "later" activities, such as systems design and even coding, begin. And it assumes that some activities may have to be redone because of errors or changing circumstances.

The top-down method of implementing a system has additional benefits as well, benefits that are in stark contrast to the conventional bottom-up approach. As we will see in Section 2.5, it encourages the developer to exercise major system interfaces early in the project; it provides "skeleton," or partially complete, versions of the system for demonstration to the user; and it permits some form of system testing and integration to take place throughout much of the implementation phase of the project.

1.4 RECENT DEVELOPMENTS

One of the "constants" of the computer field is that of technological change. Since the very beginning of the computer industry—and for virtually the entire career of any programmer, systems analyst, or project manager in the computer field—there have been steady improvements of 10 to 15 percent (or more) each year in the speed, power, size, price, and other dimensions of computer hardware; indeed, most of us have personally experienced at least two orders of magnitude—a factor of 100—in the price or performance ratio of computers that we work with. These improvements—which are certain to continue well into the 1990s, and probably well into the next century—affect project managers in at least two different ways.

The most obvious effect is that of project scale: Because computers are larger, cheaper and more powerful, organizations can tackle larger projects that affect larger (and more strategic) parts of the organization. For most organizations, this represents a fundamental shift in focus, away from the accounting-oriented "data processing" applications, and toward the executive-level "decision support" applications. The nature of this shift is beyond the scope of this book but is discussed extensively elsewhere; see Yourdon (1986) and Martin (1984) for more details.

Ever-improving computer technology has had a second major impact on project management, and this is more relevant to the subject of this book. From the viewpoint of the project manager, three technological developments have been of the greatest importance:

- Personal computers and user-developed systems
- Prototyping tools and Fourth-generation languages
- Workstations and computer-supported productivity tools

1.4.1 Personal Computers and User-Developed Systems

The advent of personal computers and user-developed systems should not change the way that a project manager goes about his job, except that (a) he won't have as many small projects to manage, because the users will do those by themselves, and (b) because of their success with small projects, users are becoming more demanding, and less sympathetic about the problems experienced by the project manager on larger projects, and (c) the project manager is now beginning to inherit the ongoing maintenance and expansion of poorly designed and poorly implemented user-developed systems.

1.4.2 Prototyping and Fourth-Generation-Language Tools

A second technological development has been the advent of prototyping tools and fourth-generation programming languages. Fourth-generation languages, such as FOCUS, RAMIS, and NOMAD, have been heralded by many experts in the computer field because they promise "factor of 10" improvements in productivity. However, since the programming phase of a typical systems development project usually accounts for only 15 to 20 percent of the total time and resources, it is not clear that even an order of magnitude improvement will have much of an overall effect.

The real impact of the fourth-generation programming languages, in my opinion, is their facility for prototyping, for developing a *working* software model of the user's new system, instead of the *paper* model that is discussed in subsequent chapters of this book.

Prototyping tools do not require the project manager to abandon the structured techniques or many of the concepts discussed in this book; however, they do change the manner in which some of the early activities of a project take place. We will discuss this in more detail in Chapter 3.

1.4.3 Workstations and Computer-Supported Productivity Tools

Much of the work of systems analysis, systems design, and computer programming requires a great deal of *manual* effort; this is ironic for an industry devoted to automating the day-to-day activities of its clients!

A number of automated tools have been developed for the *programming* phase of a project during the past 20 years—"librarian" programs, source code debuggers, higher-level languages, code generators, and the like. But the work of systems analysis and design—particularly the graphic modeling work of data flow diagrams, structure charts, and so on—remains manual in many organizations. However, now appearing in the marketplace are a number of PC-based workstations that provide automated facilities for the systems analyst and the systems designers—and for the project manager, too.

This is important for the project manager because (a) systems analysis and systems design are more time-consuming than programming, (b) the errors (bugs) associated with systems analysis and design are more serious and costly to fix than the errors of programming, and (c) truly effective maintenance of systems requires accurate documentation of the system's architecture and functional requirements, which is usually impossible with manual methods.

As this edition of the book was being prepared, it was estimated that fewer than 1 percent of the professional systems analysts in the United States had such an automated tool—an "analyst's workbench"—available to them. It is estimated that by 1990, approximately 10 percent of the professional analysts will have such technology available and by 1995 fully 50 percent will be using such tools. Because of this slow proliferation of automated tools throughout the industry, the rest of this book is written on the assumption that the project manager will *not* have such tools (other than programmer productivity tools, which are now widespread).

1.5 SUMMARY

In order to assure that you keep this entire discussion in the proper perspective, let me conclude by stating that I do not wish to imply that every project developed before the "invention" of the structured techniques was a failure; nor am I suggesting that your entire career as a project manager is worthless because you carried out projects in the conventional way. Nothing of the sort! All of your experience, and all your instincts, hunches, and subtle ways of inspiring people, soothing the anxieties of management, and disarming hostile users are still valid; if anything, those skills are every bit as important as they were years ago. What I am suggesting is that the game of developing systems has become more complex than it was even ten years ago.

Today there are more users, and those users may be more hostile to the introduction of data processing methods than they were before. (I have even seen situations in which the users can't agree about what they want among themselves, let alone with the technicians.) Many, many more programmers and analysts are collectively trying to produce 10 or even 100 times more code than on those early projects completed in the 1960s and 1970s. In short, the structured techniques are not so much a repudiation of the past and all we learned from it but rather a response to our realization that the future is almost overwhelmingly complex. We've learned from past experiences, and we know that we can be much more sophisticated about the way in which we design and develop present and future EDP systems.

The next chapter contains a brief overview of the structured techniques, intended to give you a "reading" familiarity with such approaches as structured analysis and structured design if you've never used them before. If you're familiar with the basic technical concepts of the structured techniques, you may wish to skip ahead to Chapter 3 for a discussion of the structured project life cycle. As you will see, the structured techniques are simply methods, or procedures, for carrying out the activities in the system for building systems. *That* is the system that you, in your capacity as a project manager, must be concerned with and the system to which we will devote most of the remainder of this book.

2

Structured Techniques

2.1 INTRODUCTION

It is difficult, in one chapter, to discuss half a dozen technical topics that have each been the subject of several entire books. There are, however, two legitimate reasons for my providing this overview chapter. The first is that I suspect some readers may not be familiar with one or more of the structured techniques. The brief discussion in this chapter may whet your appetite and inspire you to do some serious reading; I encourage you to choose applicable material from the references listed in the bibliography at the end of the book.

My second reason for including this brief overview is that some readers may use a slightly different definition than I use for some of the structured techniques or may apply a slightly different set of "buzzwords." This chapter should provide a reasonably good idea of the definitions and key words that I'll be using so that the discussion in the subsequent chapters will make sense.

The structured techniques that we'll be discussing in this chapter have been referred to in some organizations as "improved productivity techniques," "programming productivity techniques," or "software engineering techniques." Specifically, they include the following:

- Structured analysis
- Structured design
- Structured programming

- Top-down development
- Programming teams
- Structured walkthroughs

Each of these techniques is discussed in this chapter.

2.2 STRUCTURED ANALYSIS

One of the most exciting of the structured techniques—and potentially the most important—is *structured analysis.* As the name implies, it is concerned with the "front end" of a systems development project, during the time when the user's requirements are defined and documented.

Why is structured analysis so important? Simply because the alternative, "classical" systems analysis, is so bad. The problem is very simple to state: On anything other than a trivial systems development project, the user does not have a good advance understanding of what the data processing staff is going to develop for him. Indeed, in some cases the user has *no* idea of what the data processing people are doing for him (or to him!).

This realization comes as a terrible surprise to many systems analysts, who complain, "But we gave the user a detailed functional specification, and he signed off on it!" Most systems analysts sincerely believe that this process of "signing off" on a 1000-page functional specification will guarantee that the user knows what kind of computer system he is getting. Unfortunately, it just isn't so. Most users will tell you that they never read functional specifications in detail because the document is too long and tedious. Even if they do read the specification, they usually don't understand it.

So what's wrong with the classical functional specification? An excellent and comprehensive treatment of the subject is contained in DeMarco's *Structured Analysis and Systems Specification* (1978). But for the purposes of this overview, the problems can be summarized as follows:

- *Classical functional specifications are monolithic.* You have to read them from beginning to end—which is one reason why the number of errors in a classical functional specification increase dramatically toward the end of the document!
- *Classical functional specifications are redundant.* The same information is often presented over and over (sometimes in exactly the same way, sometimes in slightly different ways). Sometimes the redundancy is unintentional; sometimes it is a conscious effort on the part of the specification authors to create a larger (and thus more impressive) docu-

ment; sometimes the redundancy is caused by the author's effort to use the specification as a training device to "educate" the user.

- *Classical functional specifications are difficult to modify and maintain.* This difficulty is related to the first two problems. A simple change in the user's requirements often causes changes in other parts of the functional specification.
- *Classical functional specifications make too many assumptions about implementation details.* The functional specification often describes the user's requirements in very physical terms—for example, the kind of computer hardware and the physical file structure (or vendor-supplied database management system) what will be used to implement the system. These details of *how* requirements will be met have no business appearing in a functional specification and often muddle the discussion about *what* the user wants his system to do.

There is another problem that needs to be emphasized: Because in most organizations the specification is produced using primitive techniques (such as a typewriter rather than a word processor) and because of the monolithic nature of the specification (all of the 1000 pages of the specification have to be typed before any of them can be reproduced and distributed), the feedback loop between user and systems analyst is typically very long, often weeks or months. Thus if the user *does* read the specification and *does* find something that he wants to change, it will often take the systems analysis group such a long time to make the change that the user is likely to have forgotten what he wanted to change. As a result, the specification becomes perpetually out of date from an early point in the project.

All these factors contribute to make the conventional systems analysis phase of most large projects painful and time-consuming. In many cases, everyone is desperate to get the ordeal over with as soon as possible. Having finished the systems analysis phase, few wish to consider going back to reexamine or revise the functional specifications.

What, then does strucured analysis offer as a remedy to these ills? Basically, structured analysis introduces the use of *graphic* documentation tools to produce a new, different kind of functional specification—a *structured specification*—in contrast to the classical, monolithic, Victorian novel. The documentation tools of structured analysis consist of the following:

- Data flow diagrams (DFDs)
- Data dictionary (DD)
- Entity relationship diagrams (ERDs)
- State transition diagrams (STDs)
- Process specifications

The *data flow diagram* provides an easy, graphic means of modeling the flow of data through a system—any system, whether it be manual, automated, or a mixture of both. Figure 2.1 shows the basic elements of a data flow diagram: terminators (sometimes known as "sources" or "sinks"), data flows, processes, and data stores.

A typical system requires several *levels* of data flow diagrams. For example, an overview of the system might be provided in a data flow diagram such as the one shown in Figure 2.2. Each of the processes, depicted as circles or "bubbles" and shown in Figure 2.2, can be defined in terms of its own

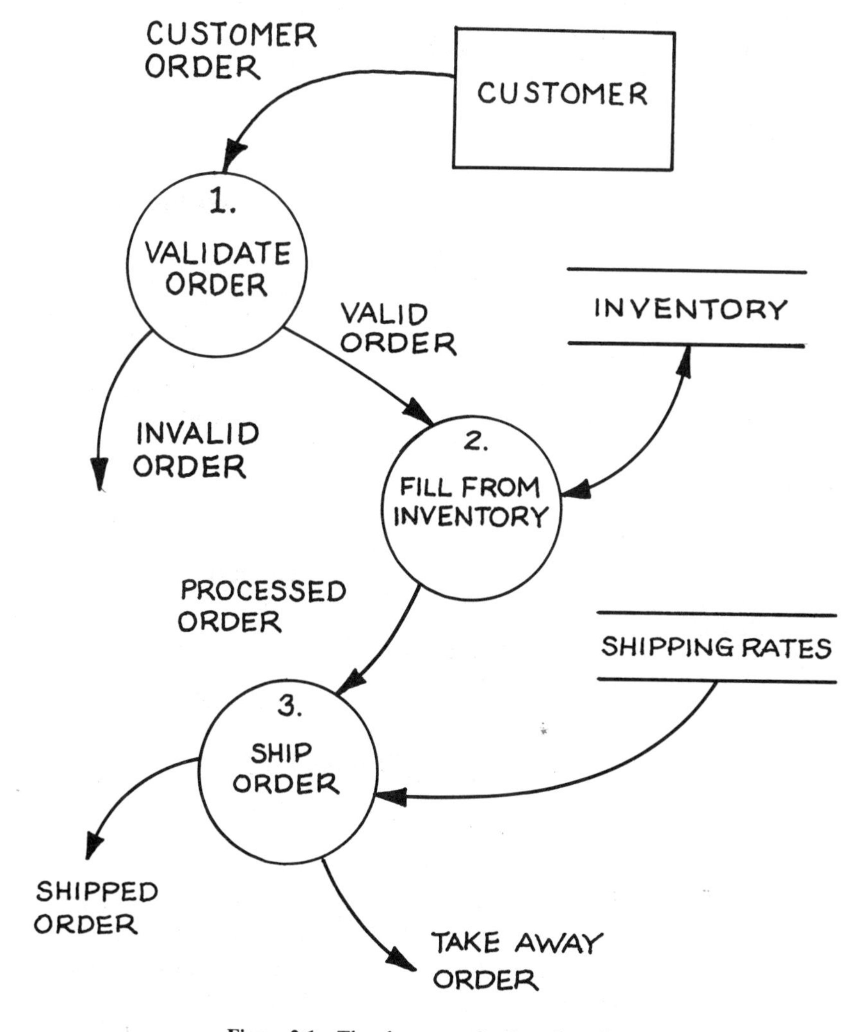

Figure 2.1 The elements of a data flow diagram.

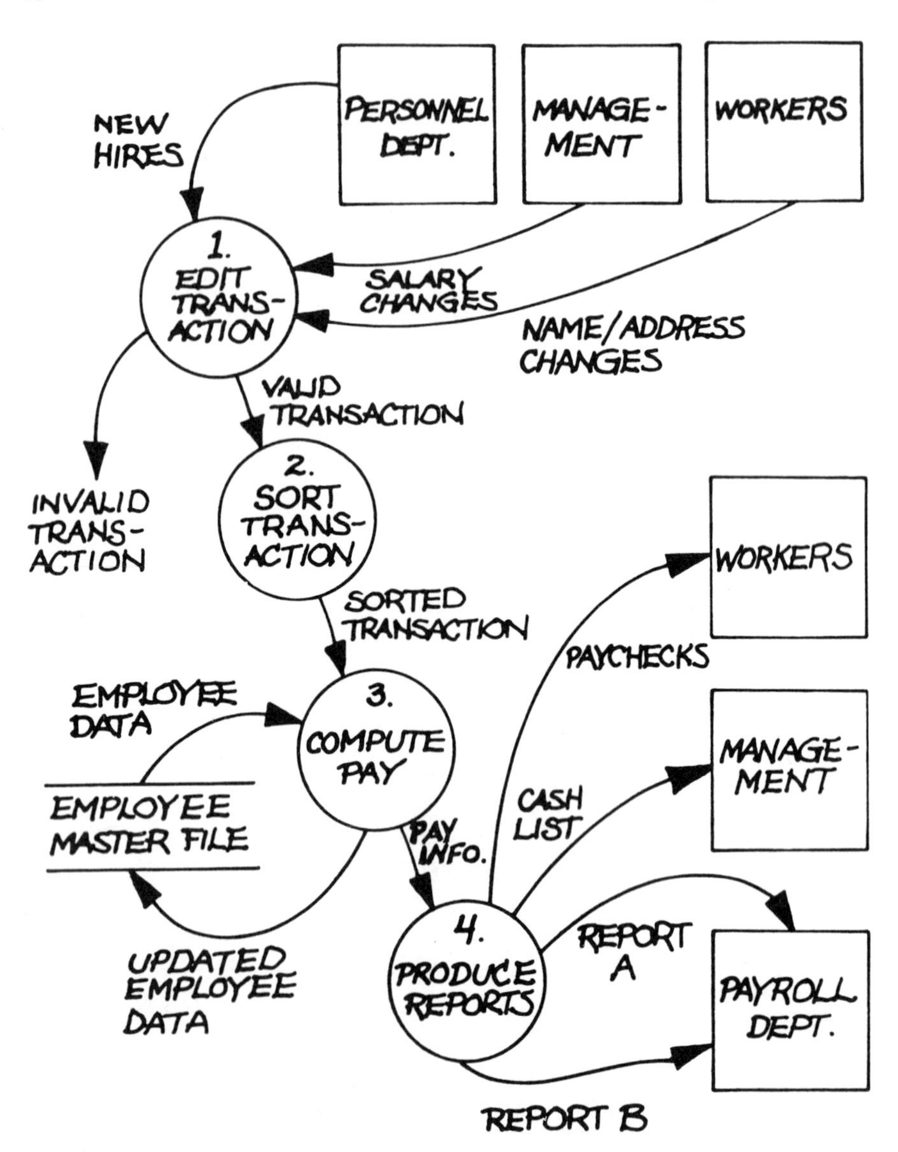

Figure 2.2 An overview data flow diagram.

data flow diagram. For example, process 3 in Figure 2.2 can be "exploded"—decomposed, or examined in more detail—in the data flow diagram shown in Figure 2.3. Note that Figure 2.3 is an exact replacement for its "parent." It has the same net inputs and outputs; it simply provides more detail.

The second major modeling tool of structured analysis is the *data dic-*

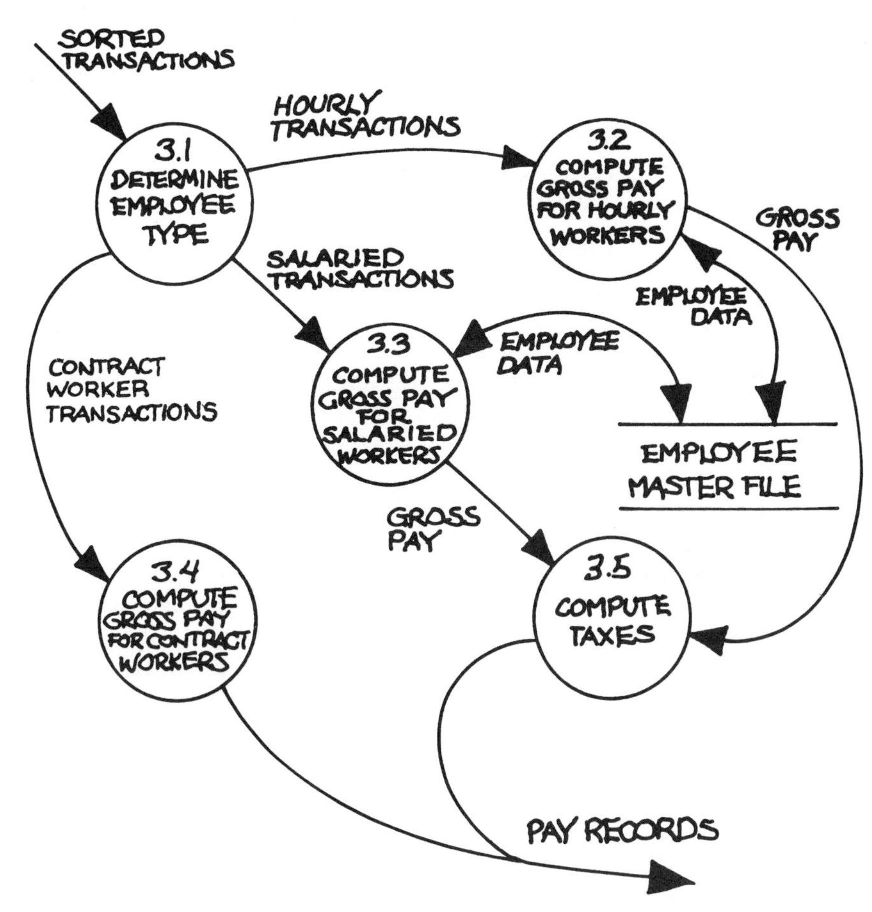

Figure 2.3 A detailed data flow diagram.

tionary, which provides supporting textual information to supplement the information shown on the DFD. A data dictionary is simply an organized collection of definitions of all the data elements in the system being modeled. In a typical system, there are likely to be several thousand such data definitions; for example, the data element CUSTOMER-ORDER shown on the data flow diagram in Figure 2.1 might be defined as follows:

```
CUSTOMER-ORDER =    CUSTOMER-NAME +
                    ACCOUNT-NUMBER +
                    [SHIPPING ADDRESS | "TAKE-AWAY"] +
                    (SALESPERSON) +
                    {ITEM-ORDER}
```

For the most part, this data definition can be read without any further supporting explanation. The notations +, [ ], ( ), and { } have precise meanings:

NOTATION	MEANING
x = a+b	*x* consists of *a* and *b*
x = [a|b]	*x* consists of either *a* or *b*
x = a + (b)	*x* consists of *a* and an optional *b*
x = {a}	*x* consists of zero or more occurrences of *a*

Thus the data element description in our data dictionary example tells us that a CUSTOMER-ORDER consists of a CUSTOMER-NAME, together with an ACCOUNT-NUMBER, together with either a SHIPPING-ADDRESS or the literal text string "TAKE-AWAY", together with an optional SALESMAN, together with zero or more instances of an ITEM-ORDER. Thus the data dictionary tells us the composition, the structure, and the meaning of the data elements in the system.

Just as with a data flow diagram, the data dictionary can present a top-down definition of a complex data element. For example, having defined CUSTOMER-ORDER, it behooves us to provide an appropriate description of its component data elements. Thus the data dictionary should also contain a definition for ITEM-ORDER:

```
ITEM-ORDER =   PART-NUMBER + (PART-NAME) +
               QUANTITY + UNIT-PRICE + (DISCOUNT)
```

Indeed, a proper job of systems analysis requires that every data element eventually be defined down to the lowest appropriate level of detail. At the lowest level, we would typically define the range of permissible values, as well as the units of measure (inches, pounds, etc.) that the data element could take on. For example, we might define QUANTITY carefully to indicate that it has a broader range, and perhaps a different unit of measure (e.g., furlongs per fortnight instead of miles per hour) than might otherwise be assumed.

The third major element of structured analysis is the *entity relationship diagram,* or ERD.[1] It highlights the major objects, or "entities," that the system deals with, as well as the relationship between the objects. The objects typically correspond, one to one, to the data stores shown on the DFD, but the DFD tells us nothing about the relationships between the objects.

[1]This replaces the "data structure diagram" found in such early structured analysis textbooks as DeMarco's *Structured Analysis and Systems Specification* (1978) and Gane and Sarson's *Structured Systems Analysis: Tools and Techniques* (1979). Detailed discussions of the entity relationship notation may be found in works by Flavin (1981), McMenamin and Palmer (1984), Ward (1984), and Ward and Mellor (1985).

An example of an ERD is shown in Figure 2.4. It shows us that there are three objects, or entities, that are meaningful to the system:

- *Employees,* about which we know such things as name, address, salary, etc.
- *Departments,* about which we know such things as location, name, budget, etc.
- *Projects,* about which we know such things as deadline, budget, manager, etc.

The ERD diagram also highlights three distinct relationships that exist between the objects:

- Employees belong to departments.
- Employees are assigned to projects.
- Departments initiate projects.

As with the DFD, the ERD requires supporting textual documentation to fill in the details: Each object and each relationship should be defined in the data dictionary. This will allow us to answer such questions as these:

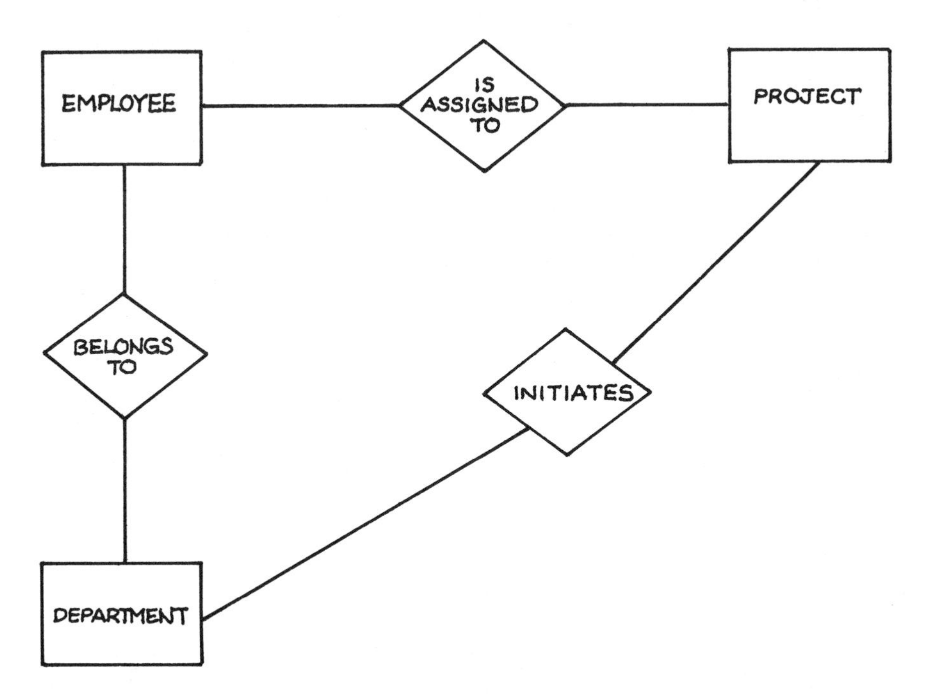

Figure 2.4.

- What does "belong" to really mean? Is the relationship one to one, or can an employee belong to more than one department?
- What does "assigned to" really mean? Can a project have more than one employee? Can an employee be assigned to more than one project?
- What does "initiate" really mean? Can a project be initiated by more than one department? Can a department initiate more than one project?

The ERD provides a simple, graphic view of the system to certain users who may not care much about the processing (or functional) details of the system. It highlights information that is not obvious in the DFD. And it provides input to the database designers who must eventually decide how best to implement the storage requirements of the system using available hardware and software resources.

The fourth major modeling tool of structured analysis is the *state transition diagram*. We need this tool because a major aspect of many complex systems is their time-dependent behavior—i.e., the sequence in which data will be accessed and functions will be performed. For some business computer systems, this is not an important aspect to highlight, since the sequence is essentially trivial. Thus, in many batch computer systems (those which are neither on-line nor real-time), function N cannot carry out its work until it receives its required input; and its input is produced as an output of function N – 1; etc.

However, many on-line systems and real-time systems—both in the business area and in the scientific/engineering area—have complex timing relationships that must be modeled just as carefully as the modeling of functions and data relationships. Many real-time systems, for example, must respond within a very brief period of time—perhaps only a few microseconds—to certain inputs that arrive from the external environment. And they must be prepared for various combinations and sequences of inputs, to which appropriate responses must be made.

The state-transition diagram (sometimes abbreviated as STD) is used to model this behavior. A typical diagram is shown in Figure 2.5 below; it models the behavior of a computer-controlled washing machine. In this diagram, the rectangular boxes represent *states* that the system can be in—i.e, recognizable "scenarios" or "situations." Each state thus represents a period of time during which the system exhibits some observable behavior; the arrows connecting each rectangular box show the *state-change,* or transitions from one state to another. Associated with each state-change is one or more *conditions* (the events or circumstances that caused the change of state), and zero or more *actions* (the response, or output, or activity that takes place as part of the change of state.

The fifth major element of structured analysis is the *process specification.* Its purpose is to allow the systems analyst to describe, rigorously and precisely, the *business policy* represented by each of the bottom-level "atomic" processes

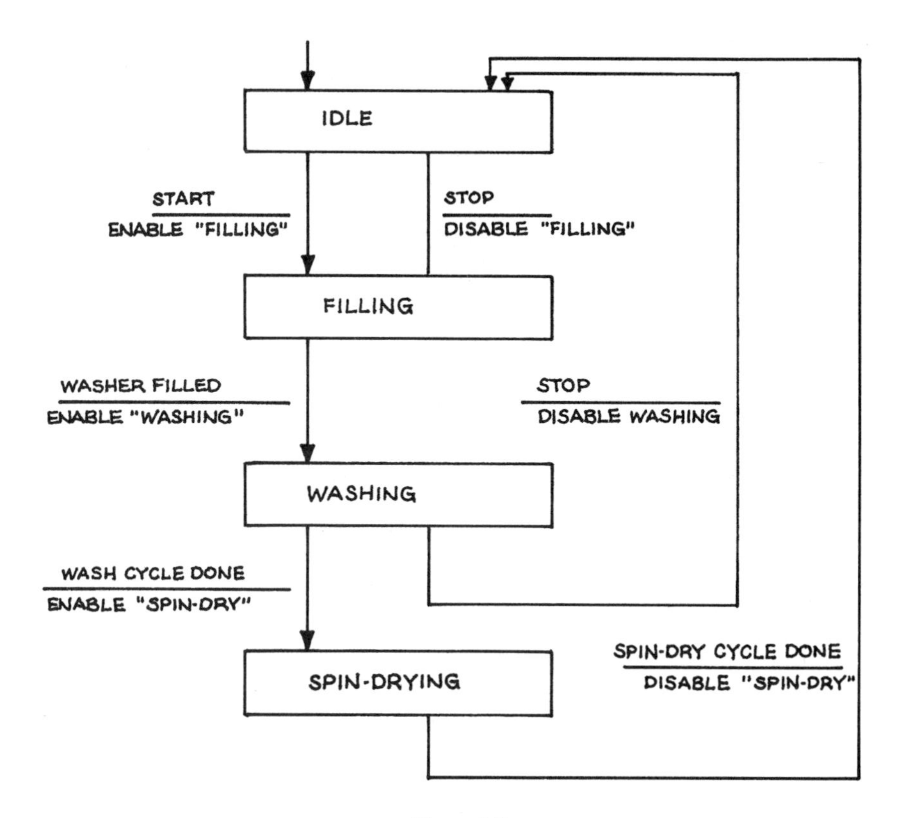

Figure 2.5.

in the bottom-level data flow diagrams. Process specifications can be written in a variety of ways: decision tables, flowcharts, graphs, "pre" and "post" conditions,[2] and structured English. At the present time, structured English appears to be the most commonly used form of process specification; here is an example of a structured English process specification (or "minispec") for a process called CARRY OUT BACK-BILLING:

1. IF the dollar amount of the invoice times the number of weeks overdue is greater than $10,000 THEN:
   a. Tell the salesperson to call the customer.
   b. Examine the invoice again in two weeks.

[2]"Pre" and "post" conditions are ways of describing, respectively, the conditions that must be true before the process operates and the conditions that must be true after the process operates. The algorithm, or procedure, by which the process produces its output is not described, thus leaving it to the designer and programmer to choose the most appropriate algorithm.

2. OTHERWISE IF more than four overdue notices have been sent THEN:
   a. Tell the salesperson to call the customer.
   b. Examine the invoice again in one week.
3. OTHERWISE (the situation has not yet reached serious proportions):
   a. Add 1 to the overdue notice count on the back of the invoice (if no such count has been recorded, then write "overdue notice count = 1").
   b. Send the customer a photocopy of the invoice, stamped "Nth notice: invoice overdue. Please remit immediately," where N is the value of the overdue notice count.
   c. Log on the back of the invoice the date on which the Nth overdue notice was sent.
   d. Examine the invoice again in two weeks.

In its most extreme form, structured English consists of only the following:

- A limited set of imperative verbs, such as *compute* and *validate*—perhaps as few as 40 or 50
- Control constructs borrowed from structured programming (such as IF-THEN-ELSE and DO-WHILE) to describe alternative actions and repetitive actions
- Nouns that have been defined in the data dictionary or are defined "locally" within the process specification itself

Various other methods can be used to document business policy for bottom-level processes. Depending on the nature of the application, decision tables, mathematical formulas, or graphs—or a mixture of such techniques—might be more appropriate (indeed, most complex systems development projects do require a combination of documentation techniques for the process specifications). The key point, though, is that each process specification stands on its own and describes the business policy for only one small piece of the overall system.

We have now seen the major pieces of structured analysis: data flow diagrams, data dictionaries, entity relationship diagrams, state-transition diagrams, and process specifications. Even though the description has been brief, you should be able to see that a structured specification has a number of desirable characteristics, summarized as follows:

- It is *partitioned* rather than being a monolithic entity.
- It is *graphic,* consisting largely of pictures rather than words.
- It is *top-down,* presenting a description of a system at progressive levels of detail.

- It is *implementation-independent,* presenting an abstract model of the system that will be developed for the user.

Although this brief overview is not intended to enable you or your systems analysts to begin practicing structured analysis with any degree of proficiency, at the very least it should provide you with enough information to get started. For more detail, consult the sources on structured analysis cited in the bibliography.

2.3 STRUCTURED DESIGN

The second structured technique that we will review briefly is known as *structured design.* Although it is based on work carried out in the later 1960s and early 1970s, it first attracted widespread attention with the publication of a now-classic article titled "Structured Design."[3]

Perhaps the most important point to understand is that structured design is not just modular programming, or modular design, as it was called in the late 1960s, nor is it just "top-down design." Modular programming was a wonderful idea, but nobody ever defined accurately what was meant by a module, nor did anyone ever distinguish precisely between a "good" module and a "bad" module. Top-down design is a good idea, too, but by itself it's not sufficient: It's possible to design a terrible system from the top down.

Structured design addresses these issues directly and can be defined as "*the determination of which modules, interconnected in which way, will best solve some well-defined problem.*" Note that we assume in this definition that we have a well-defined problem; after all, a brilliant solution to the wrong problem won't do us much good. So we're assuming that structured design has been preceded by structured analysis. Notice also that structured design is defined as the process of determining the *best* solution to a problem. This brings up an interesting philosophical problem: Is there such a thing as the *single* best solution to a data processing problem? If so, would we recognize it if we saw it? Would other designers agree that it's the best design? And do we have a systematic method that will assure us of generating the single best design in a finite amount of time?

Thinking about design in these terms was suggested by an architect, Christopher Alexander, in a delightful book titled *Notes on the Synthesis of Form.*[4] In the book, Alexander points out that designers—whether airplane

[3]W. Stevens, G. Myers, and L. Constantine, "Structured Design," *IBM Systems Journal,* 13, no. 2 (May 1974), 115–139.

[4]Christopher Alexander, *Notes on the Synthesis of Form,* 2nd ed. (Cambridge, Mass: Harvard University Press, 1971).

designers or software designers—have a difficult time agreeing on the characteristics of a "good" design. However, designers are very good at spotting "bad" design, or "instances of misfit," as Alexander calls it, and they can generally agree about the existence of badness in a design. So why not define "good" design as the absence of easily recognizable badness?

In the software field, we have to assume that there are an infinite number of designs for anything other than a trivial problem. Some of the designs don't work at all; some are terrible, without any redeeming features; some are mediocre; some are tolerable; and a few are very good. Although we can probably never isolate the single best design, at least we should be able to eliminate the large class of bad designs, and if we end up arbitrarily selecting one design from the (presumably) smaller class of good designs, we will have improved the state of the art considerably.

There are two other philosophical points to be considered before we discuss the specifics of structured design. What do we mean by "good?" And what do we mean by "module"? If you ask your programmers and systems analysts to write a short essay on each of these two questions, you will be amazed by the results.

You'll find, for example, that many programmers and systems analysts still equate "goodness" with hardware efficiency; they'll state, "A good design is one that doesn't waste too many microseconds of CPU time." Structured design takes a much more pragmatic view: A good design is a *cheap* design, one that is inexpensive to develop, inexpensive to operate, and inexpensive to modify and maintain. Since operational costs (the cost of computer hardware) have fallen steadily for the past three decades, with no end in sight, and since software maintenance[5] now represents more than 50 percent of the lifetime cost of a system, structured design particularly emphasizes *maintainable* systems.

You'll also find that many programmers and systems analysts define the characteristics of a module in terms of specific hardware features or programming language features, declaring perhaps that "a module is a COBOL paragraph that doesn't take more than 4096 bytes of computer memory." Or they'll come up with a "religious" definition, such as "a module is something that fits onto one page of a program listing, takes no more than 50 statements, is well documented, and has no more than three levels of nested IF-THEN-ELSE statements." Structured design insists on a simple, nonreligious, non-vendor-oriented definition of a module: A module is a contiguous, bounded group of program statements having a single identifier (name) by which it can be referenced as a unit.

[5]In this context, software maintenance comprises the activity of fixing a system when it operates incorrectly and also the activity of modifying a system when the user's requirements change or when the underlying hardware or software environment changes.

Having discussed the philosophy behind the development of the concepts of structured design, we can now discuss the components of structured design:

- Documentation techniques
- Design evaluation criteria
- Design heuristics
- Design strategies

2.3.1 Documentation Techniques

As with structured analysis, the documentation techniques associated with structured design include both graphical tools and textual information. The graphical documentation tools of structured design include data flow diagrams, HIPO diagrams, and structure charts; the textual information con-

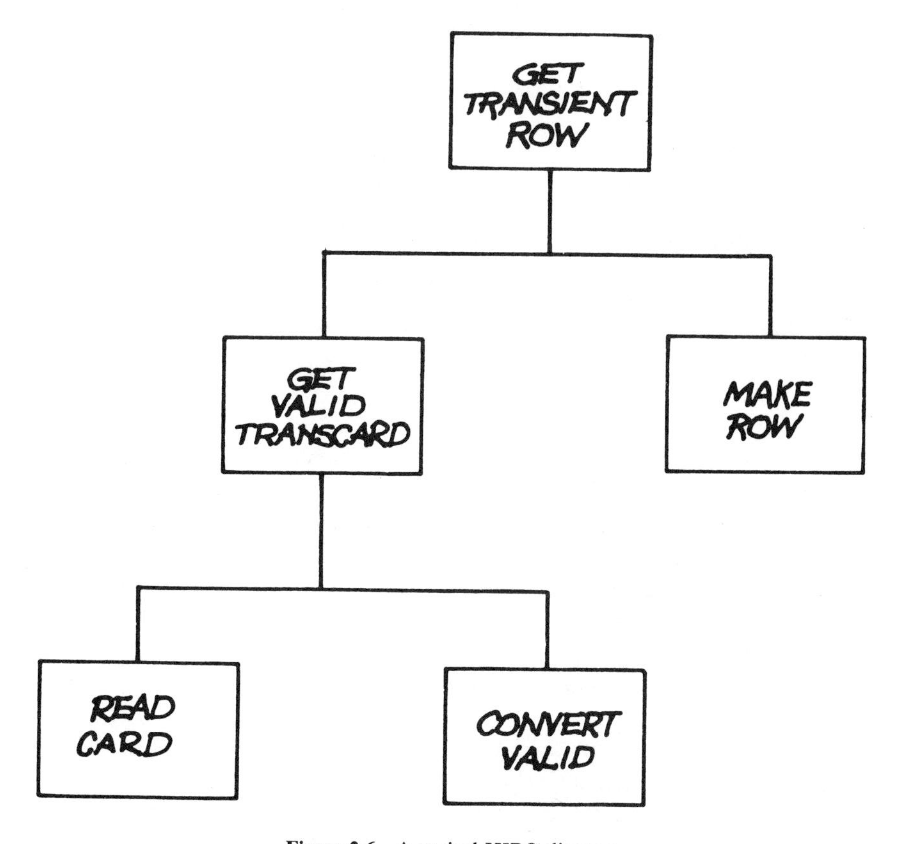

Figure 2.6 A typical HIPO diagram.

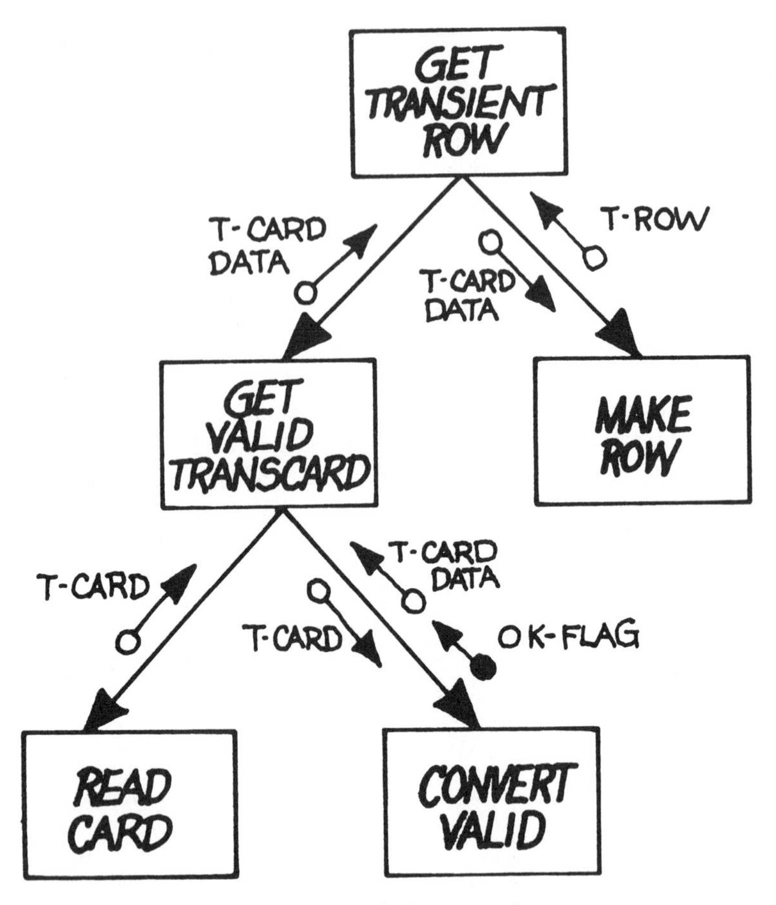

Figure 2.7 A typical structure chart.

sists of module specifications and a data dictionary. The purpose of all these documentation tools is to help model the *structure* and *architecture* of a system of modules, rather than concentrating on the procedural logic. Figure 2.6 shows a typical HIPO diagram; Figure 2.7 shows a structure chart of the same system. As you can see, the diagrams are similar, although structure charts contain more information than HIPO diagrams.

2.3.2 Design Evaluation Criteria

The real core of structured design is *design evaluation criteria:* objective methods of evaluating a design, as documented in a structure chart or HIPO diagram, to see whether it is likely to be a good design or a bad design. As we mentioned earlier, it's almost impossible to get everyone to agree about goodness, but it's relatively easy to spot instances of badness. Two concepts, known as *coupling* and *cohesion,* help us do just that.

Coupling is a way of measuring the strength of interconnections between one module and another. Coupling is thus a "bad" characteristic, since in a system with strongly coupled modules it will be almost impossible to maintain or modify one module without making modifications to other modules. More important, coupling can be used as a relative measure; for example, suppose that two different designers developed the designs shown in Figures 2.8 and 2.9 from the same specifications. As we can see, the modules in design 1

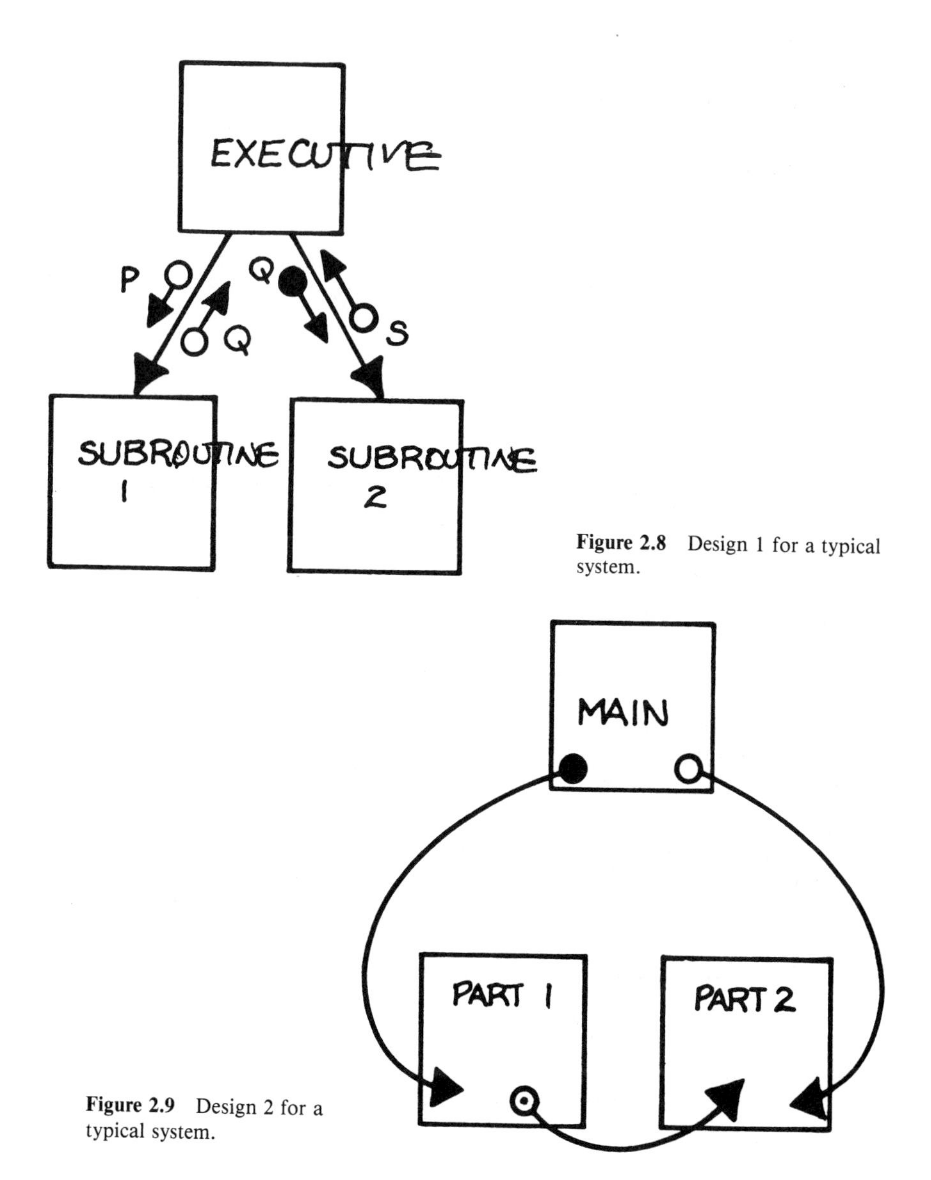

Figure 2.8 Design 1 for a typical system.

Figure 2.9 Design 2 for a typical system.

are connected by a simple subroutine call (or, in COBOL, a CALL USING statement), with one simple input parameter and one simple output parameter. In design 2, the two modules have various "pathological connections" to one another—that is, one module refers to, or even modifies, the inside of another module. Clearly, any modification of one module in design 2 will require careful study, and possibly some modifications, of the other module, and this is obviously bad. There are a number of detailed guidelines in structured design for evaluating the coupling between modules; these guidelines are not discussed in detail here but can be found in such textbooks as Yourdon and Constantine's *Structured Design* (1979) and Page-Jones's *Practical Guide to Structured Systems Design* (1980).

Cohesion is, in a sense, the opposite of coupling: It is a measure of the strength of connection of elements (that is, individual program statements of subordinate modules) *inside* a module. In other words, describing a module's cohesion is a way of describing the degree to which the module carries out a single, well-defined job; in the negative sense, it is a way of describing the randomness or unrelatedness of elements contained within a module. For example, the modules shown in Figure 2.10 have either very low cohesion or

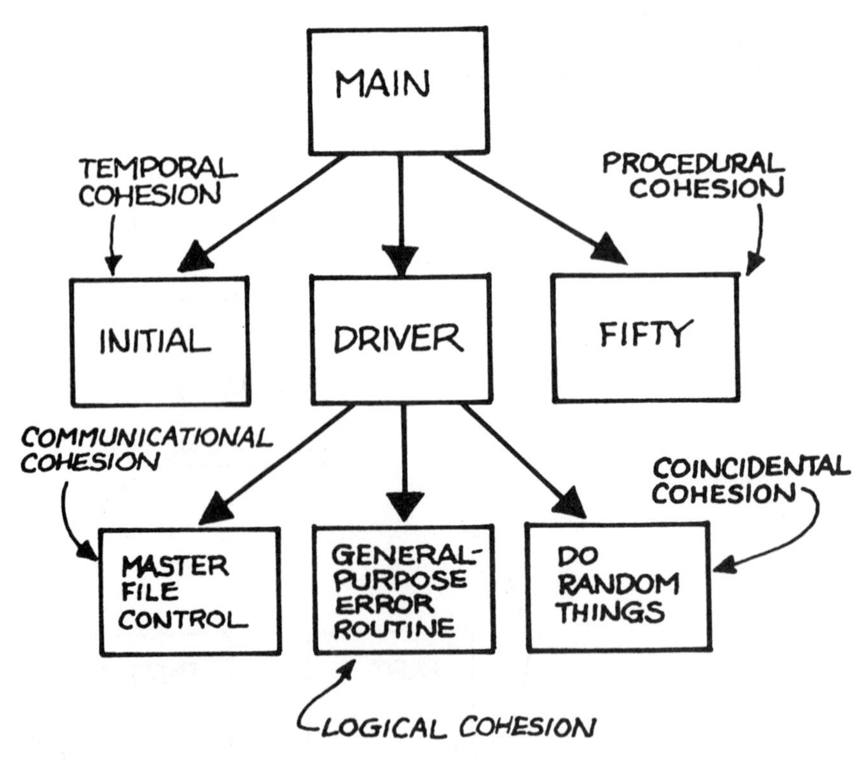

Figure 2.10 Examples of cohesion.

dubious cohesion—the module called GENERAL-PURPOSE-ERROR-ROUTINE, for example, contains elements that, instead of carrying out one single, well-defined function, are engaged in several distinct (though similar) functions; such a module is said to have *logical cohesion.*

There are various other levels of cohesion: coincidental, temporal, procedural, communicational, sequential, and functional. The "good" level of cohesion is *functional* cohesion. Most levels of cohesion can be thought of as easily recognizable examples of badness (an example of temporal cohesion is INITIALIZE; an example of communicational cohesion is MASTER-FILE-CONTROL). Odd though it may sound, functional cohesion can be thought of as the absence of all other bad levels of cohesion; if the designer can demonstrate that his proposed design does not contain any modules with instances of bad cohesion, the modules must have functional cohesion. For those who insist on a positive way of identifying functionally cohesive modules, I suggest the following guideline: If you can describe, accurately and honestly and completely, what your module does (not *how* it does the job but simply *what* the job is) in a simple English sentence containing a single transitive verb and a single object, you probably have a functionally cohesive module.

Coupling and cohesion go hand in hand: A system whose modules have low cohesion will almost certainly have strong coupling. Conversely, if the modules have strong, functional cohesion *at every level* in the hierarchy, the coupling will be minimized and the overall quality of the design will be significantly enhanced.

2.3.3 Design Heuristics

Design heuristics are simple rules of thumb that are generally useful, but which can't be guaranteed to work in all cases. The most popular design heuristics relate to module size and control:

- *Module size.* One can generally assume that small modules are relatively simple (although some programs written in such languages as APL can be good counterexamples). By small, we mean a module that consists of 50 to 100 program statements. Since simple modules tend to be easier to maintain and modify than large (complex) modules, it follows that structured design favors small modules. Typically, the guideline indicates that the code for a module should fit onto a page or two of a program listing, so that the programmer or reviewer can lay the listing on his desk and see the entire module in front of him.
- *Span of control.* Just as a manger in a human hierarchy should not try to control too many immediate subordinates, so a "manager module" should not try to control too many immediate subordinates. Structured design suggests that the maximum a module should control is five to nine

(or "seven plus or minus two") immediate subordinates, except for so-called transaction centers.[6]

- *Scope of effect and scope of control.* The heuristic pertaining to effect and control can be explained through a management analogy: If a clerk's day-to-day activities are governed by a management decision, the clerk should be working somewhere within that manager's empire—that is, within the manager's "scope of control." In software design, we suggest that any module whose behavior is affected by a decision ought to be subordinate to the module that makes the decision.

As you can see, these guidelines are simple, but keep in mind that they are only guidelines, not rigid rules to be followed religiously at any cost.

2.3.4 Design Strategies

The last component of structured design is *design strategies.* Once again, explanation by analogy is probably the simplest. Imagine that you are a brilliant chef, with years of experience in the best restaurants in the world, and that you've just developed a craving for chocolate cake. What would you do? You'd probably go into your well-equipped kitchen and create a chocolate cake, on the spot. All steps of the preparation and cooking process would flow smoothly; there would be no hesitation about what ingredients to use, consistency, batter, or anything else.

Of course, you're probably *not* a professional chef. Faced with a craving for chocolate cake, you would react rather differently: First, you would look in a cookbook for a reasonable-looking recipe. That's a nice, simple, deterministic instruction; almost anyone can follow it, and virtually everyone can tell whether it's been done properly. Unfortunately, some of the steps in the typical recipe are a little more difficult: "Cream the butter." When exactly does butter become cream? Or "Cook until the cake shrinks from the side of the pan." What is the termination criterion on that step? How can you tell whether that step has been done correctly before it's too late?

Clearly, cookbook recipes assume a certain minimum level of experience, common sense, and judgment. Exactly the same situation is true in the field of software design. Brilliant designers don't need recipes; they can do

[6]A transaction center is a module that receives, as input, a transaction (whose type and nature can be determined by examining some kind of transaction code) and then dispatches the transaction to one of several subordinate modules for appropriate processing. Since there may be dozens (or even hundreds) of different types of transactions for a typical application, the transaction center may have dozens of immediate subordinates. But since the dispatching is based on a single, rather simple decision ("What type of transaction is this?") and invokes only a single processing path, we don't have to worry about the transaction center module being inordinately complex.

everything by instinct. Average designers find cookbook instructions extremely helpful, but even the best recipe in the world can't help an incompetent designer.

There are two popular design cookbooks in use at the present time, one based on data *flow* and the other on data *structure*. The data flow strategy begins with the data flow diagrams produced by the system analyst and follows a step-by-step procedure to produce a structure chart (or HIPO diagram); this is illustrated by Figure 2.11.

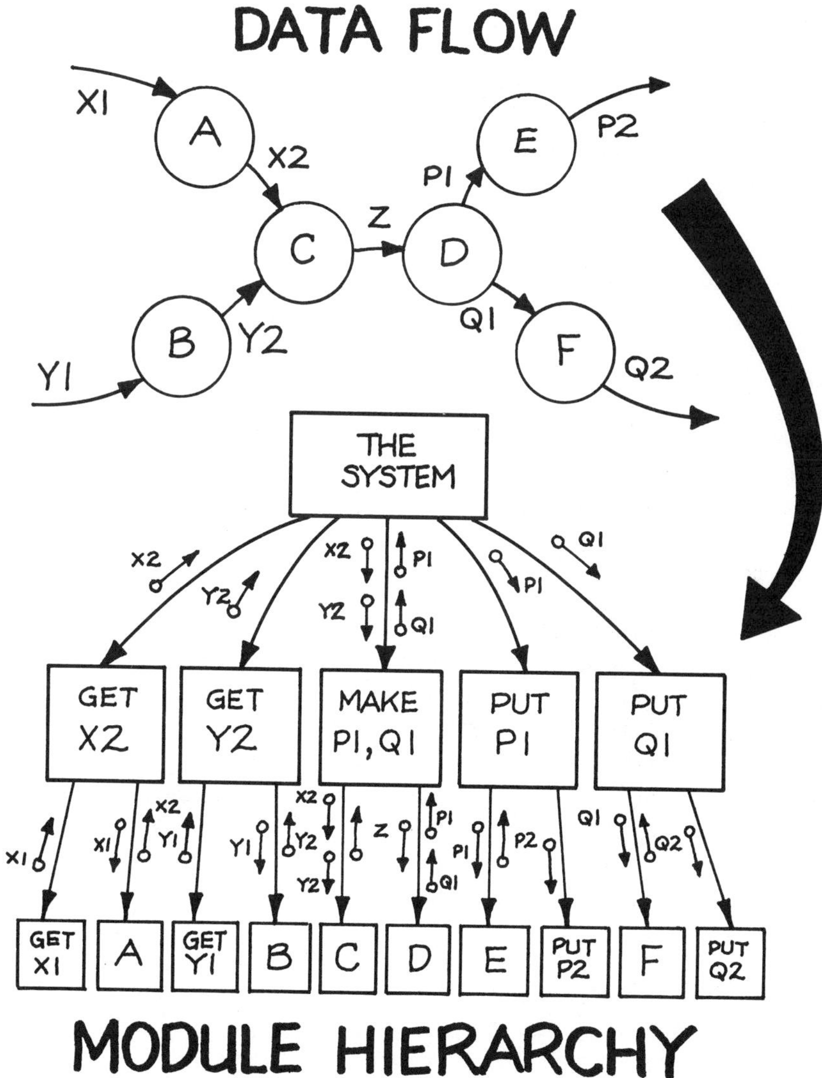

Figure 2.11 Illustration of a data flow design.

The data structure approach assumes that the designer is working with one or more simple hierarchical data structures (for example, fields within records and records within files) and uses the data structures associated with the problem as the first step in a step-by-step procedure to produce a structure chart. This is illustrated in Figure 2.12.

These design strategies, combined with the documentation techniques, evaluation criteria, and heuristics, are the elements of strucured design. And it should be emphasized, as we finish our discussion of this topic, that structured design is *all* of the topics discussed. Unfortunately, some designers seem to feel that the use of HIPO (or a structure chart) is equivalent to applying the techniques of structured design. HIPO and structure charts are good documentation tools, but their use alone does not guarantee a good design. Similarly, some designers feel that cookbook approaches are sufficient to

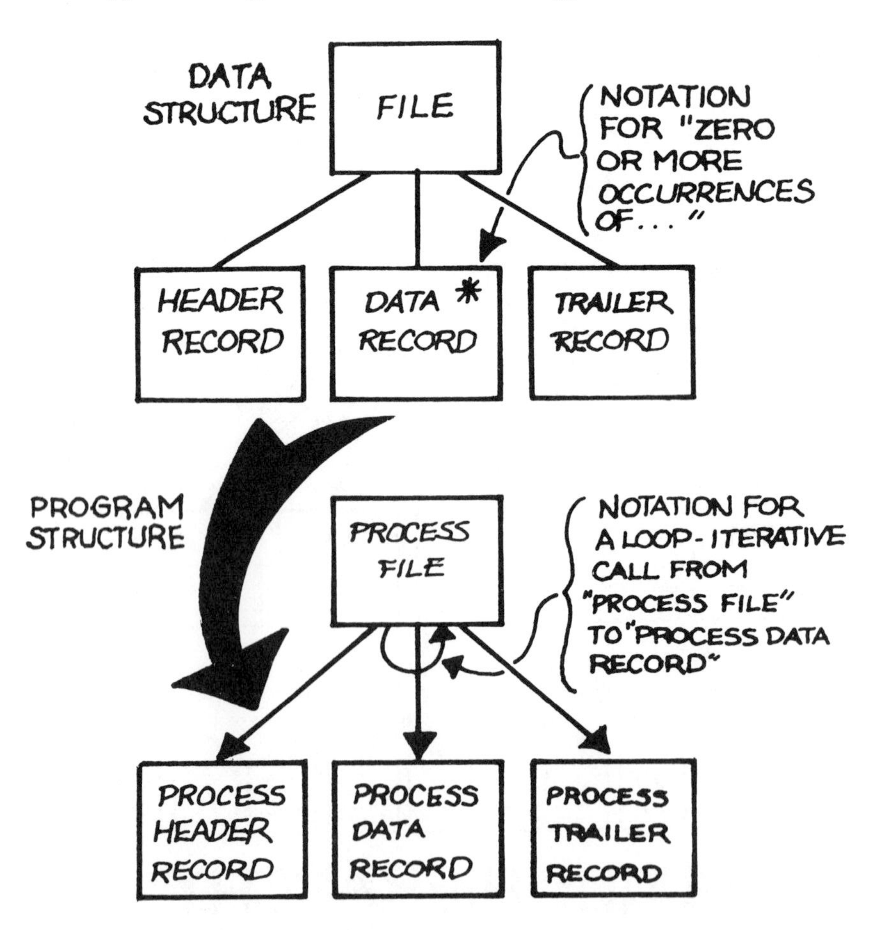

Figure 2.12 Illustration of a data structure design.

produce a good design, but without the concepts of coupling and cohesion, together with the design heuristics we discussed, cookbook recipes are inadequate.

2.4 STRUCTURED PROGRAMMING

Structured programming was the first of the structured techniques to be discussed and practiced widely. As a discipline, it is based largely on fundamental theoretical work carried out in the 1960s by such people as Edsger Dijkstra, Niklaus Wirth, and Bohm and Jacopini.

To many people, the term *structured programming* is generic and includes design philosophies, implementation strategies, concepts of program organization, and the Boy Scout virtues of thrift, loyalty, and bravery. For the purposes of this discussion, however, structured programming should be thought of as a *coding* technique and should be discussed only in the context of programming.

The theory behind structured programming is relatively simple: It states that any program logic can be constructed from combinations of the three constructs shown in Figure 2.13. In most high-level (third-generation and fourth-generation) programming languages, this means that computer programs can be constructed from the following combinations:

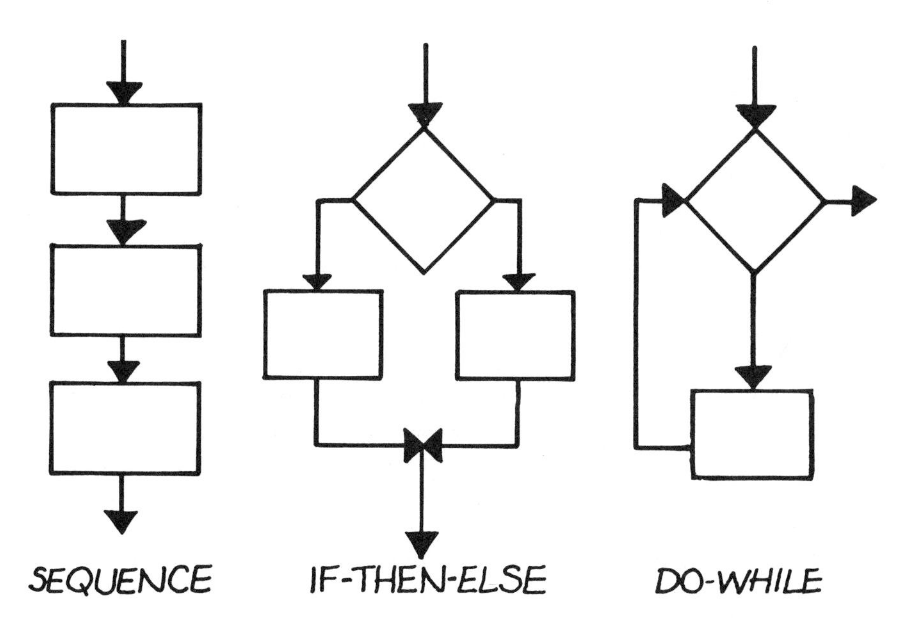

Figure 2.13 Structured programming constructs.

- Simple imperative statements, such as COMPUTE, READ, WRITE, and algebraic statements such as X = Y + Z
- IF-THEN-ELSE statements and CASE statements for decision making
- DO-WHILE and DO-UNTIL statements for iteration

The key point is that this set of constructs is *sufficient* to form any and all kinds of program logic; in particular, it is not necessary to use uncontrolled branching statements—the infamous GOTO statement—in structured programs.

I first began teaching structured programming concepts in 1971; at the time, few people were interested in structured programming, and still fewer believed that the approach would work. Virtually every one of the programmers in my courses and seminars had to be shown—usually by working on a program that he or she had already written—that it was indeed possible to rearrange the logic so that it could be expressed with DO-WHILE and IF-THEN-ELSE logic. Unfortunately, even if I was able to convince my students that structured programming was possible, I generally couldn't convince them that it was worth the effort. It was only after publicity in such journals as *Datamation* and the *IBM Systems Journal* that the idea began to catch on. Unfortunately, there was so much publicity that by the mid-1970s, structured programming began to take on some of the overtones of a religion: Programmers began to worry that a lightning bolt would strike them down if they coded a GOTO statement in the program.

Today, however, almost all of the furor has subsided. The majority of EDP professionals realize that good coding is important, but it's not a life-and-death issue. It is also generally understood that brilliant code may not save a mediocre design: If you haven't used structured design, there may not be much point in using structured programming. Furthermore, almost all students trained in university computer science programs are learning structured programming concepts (in such languages as Pascal, C, or Ada) from the beginning, so there's no need to convince them to write their programs any other way.

There are, of course, some problems that even the use of structured programming techniques may not resolve. Those problem areas pertain to the suitability of the structured techniques to the particular programming language, the skill level of the programmers in the use of their language, and the proportion of time spent on actual programming to time spent on maintenance. It is generally agreed that languages like Pascal, C, Ada, and PL/I are superior languages for writing structured code. COBOL is adequate, though its mechanism for forming decisions and loops is clumsy, and the lack of local variables and formal block structures is extremely annoying. Fortunately, the COBOL language is undergoing constant revision, and we may expect that the COBOL of the late 1980s and the 1990s will make structured programming natural and convenient. Similarly, FORTRAN and BASIC began as mediocre languages (from the viewpont of structured programming) but have been extended and improved to include the necessary structured programming constructs.

However, even if your organization is using a good programming language, there is no guarantee that your programmers are using it properly. New programmers are trained in college to write structured programs in Pascal and C, but veteran programmers who learned to program in COBOL or FORTRAN ten years ago may not have learned how to use block structures, DO-WHILE statements, or nested IF statements. COBOL programmers in many organizations are notoriously wary of nested IF-THEN-ELSE statements; they're not familiar with the PERFORM-UNTIL statement (which is the common COBOL construct for forming GOTO-less loops), and they're shy about using the CALL USING statement to pass parameters from one module (or subprogram) to another. Thus a certain amount of refresher training may be necessary to ensure that structured programming is being practiced in something other than a superficial fashion.

Finally, you should determine whether it's possible to introduce structured programming at all. If your staff spends 98 percent of its energies patching vintage 1972 code, there may not be much of an opportunity to do any creative structured programming. Maintenance programming is a serious problem; how to deal with the problem raises questions beyond the scope of this book. One possibility is to use one of the vendor-supplied restructuring "engines" to convert old, unstructured programs into equivalent structured programs. However, for the purposes of this book, I will assume that you are beginning a systems development project on which your staff will have the opportunity to write all new code.

2.5 TOP-DOWN DEVELOPMENT

Next on our list of structured techniques is *top-down development.* Actually, what I am interested in is something that might be called top-down *implementation,* in contrast to the popular concept of top-down *design.* Indeed, you should be aware that many people in the data processing field use the term *top-down* rather loosely. For reasons that are largely historical, top-down design and structured programming have often been discussed as if they were a single topic.

There is nothing particularly wrong with the fact that people have long lumped top-down design, structured programming, and structured design together. Just be sure that when you're discussing top-down development with one of your colleagues, he's not thinking of structured design. Also, make sure that you're both discussing the same *aspect* of the top-down approach. Three related but quite distinct aspects of "top-down" are defined as follows:

- *Top-down design* is a design strategy that breaks large, complex problems into smaller, less complex problems and then decomposes each of these smaller ones even further until the original problem has been expressed as some combination of many small, easily solvable problems.
- *Top-down coding* is a strategy of coding high-level (or "executive") modules as soon as they have been designed and generally before the lower-level "detail" modules have been designed.
- *Top-down implementation* (also known as *top-down testing*) is a strategy for testing the high-level modules of a system before the low-level modules have been coded and possibly before they have been designed—indeed, sometimes even before their functional requirements (as seen by the end user) have been specified.

Let me give you an example of top-down development. A few years ago I participated in a project in Sydney, Australia, to develop a payroll system. When the system was finished, it represented approximately 4 person-years of work and 30,000 COBOL statements. Even though it was obviously a medium-sized system with some inherent complexities, the project team began writing code approximately *two weeks* after the project began—and we actually had a working, demonstrable payroll system only six weeks after the beginning of the project.

Of course, our preliminary version of the payroll system had a few minor limitations: The user was required to provide error-free transactions in the proper sequence, since our payroll system would neither edit nor sort the transactions. In actuality, there wasn't much point in putting many transactions into the system, since we were unable to hire any new employees, fire any old ones, give anyone a raise, or, for that matter, make *any* changes to an employee's current status. Not only that, our payroll system uniformly paid everyone $100 per week and withheld $15 in taxes. It insisted that all employees be paid by check (instead of cash or direct bank deposit), and—the final insult!—it printed all of the paychecks in octal.

Obviously, this was a rather limited payroll system and not really fit to show the users. However, it *did* involve all of the modules at the top level, which was all we had time to design before we began coding. What made the system so primitive was the fact that all of the lower-level modules were "stubs," or "dummy routines." For example, the top-level module in the update program called a lower-level module to compute an employee's salary; for the first version of the payroll system, that module simply returned an output of $100, regardless of the employee's name, rank, or seniority. Similarly, the top-level module in the edit program called a lower-level module to validate a specific transaction; in our first version of the system, that module simply returned with an indication that the transaction was valid—without going through any effort actually to validate the transaction.

Subsequent versions of the payroll system merely involved adding lower-level modules to the existing skeleton of top-level modules. A second version of the system, for example, allowed the user to hire and fire employees; it also sorted the transactions; and, in a very few simple cases, it actually computed an employee's gross pay. However, version 2 still made no attempt to validate the input transactions, and in most cases it still paid employees $100 per week, in all cases it still withheld $15 in taxes, and, unfortunately, it still printed the paychecks in octal. Subsequent versions added in the details, until a final version produced output that was satisfactory to the user.

So that's top-down development. For many years, it has been considered a *testing* strategy—a technique that the programmers would use during the implementation phase of a project to make their life easier. However, it's not recognized that top-down development is more than just a testing strategy: It's a manifestation of the *iterative* project life cycle concept. By producing a skeleton version of a computer system, we're building something that we can show the user at a relatively early stage. And by showing a preliminary system to the user, we provide the opportunity to go back into the systems analysis phase of the project—to double-check that the specifications were correctly understood by the user and to confirm that the use hasn't changed his mind about what he wants the system to do. And we also have an opportunity to see how the skeleton version of the system actually runs on the computer hardware, a feature that may point out flaws in the design (or shortcomings in the hardware!).

This concept of iterative development of EDP systems is the foundation of much of the remainder of this book. Chapter 3 presents the iterative project life cycle in more detail, further exploring the concept of top-down development.

2.6 PROGRAMMING TEAMS

Still another aspect of the structured systems development approach is the concept of programming teams, in particular, *chief programmer teams* and *egoless teams.* In the following paragraphs, I will touch on the evolution and modern-day application of teams in the context of the data processing environment.

The chief programmer team concept is based on an experimental project that IBM first carried out in the early 1970s, but it is a concept that many EDP organizations, especially in the software consulting and computer manufacturing fields, have been practicing for decades. The chief programmer team, as the name suggests, is built around a chief *superprogrammer.* Fred Brooks describes the chief programmer as someone who has at least ten years of experience, can develop software ten times faster than the average person, is capable of doing both the systems analysis and the implementation of the software, can write all of the documentation, and can supervise other members of the team (see Brooks, 1975, for more details). As you can appreciate, there aren't very many such people around, but if you could find a few, you could accomplish miracles in your EDP organization!

Assuming that you can find a chief programmer or two, you would want to supplement their talents with a copilot, an administrator, a language lawyer, a toolsmith, and a librarian. These additional members of the team are described as follows:

- *Copilot.* This person is, in a sense, the backup chief programmer, ready to take over if the chief programmer has to leave the project for some reason.
- *Administrator.* On large projects, this person would be responsible for all of the paperwork, contractual and legal negotiations, scheduling of computer time, and the myriad of other tasks that might otherwise keep the chief programmer from carrying out technical work.
- *Language lawyer.* Even though the chief programmer is presumably bright enough to become an expert in almost anything, relying on him to know everything about everything is usually not a good use of his time and talent. Instead, the project team should include one or more language lawyers, people who are experts on the intricate details of various systems, people who have read the vendor manuals for the operating system, the programming language, or the database management system. This is the person to whom one turns when the vendor's software begins producing cryptic diagnostic messages that aren't documented anywhere or when there is an urgent need to squeeze as much efficiency as possible out of a specific portion of the system.
- *Toolsmith.* This person excels in building "software tools" or "utility programs" for use by the chief programmer and other members of the team. Examples of tools might be test data generators, special debugging programs, and programs for building and manipulating a data dictionary. In some cases the toolsmith's real talent is showing everyone else on the team how to find and use the organization's existing tools and utility programs.
- *Librarian.* This person is, in many people's opinion, the most important member of the team. He files, organizes, and updates all of the material associated with the develoment project: the data flow diagrams, data dictionary, source program modules, program listings, and object programs. Even if automated support is available (which is more and more common in EDP organizations today), a large project can easily degenerate into chaos without a human librarian: There is a great

tendency for each systems analyst and programmer to maintain his own private version of his work, without properly informing other project members of the whereabouts of his most up-to-date version.

Frankly, most organizations don't have any chance of all of developing a *real* chief programmer team. There are four major reasons for this. First, there are very, very few people in the computer field talented enough to be called chief programmers. It's ironic that almost every programmer likes to think that he is a chief programmer; on closer examination it often becomes obvious that even though he can code quickly, he can't document anything, he can't communicate with the end user, and he can't even supervise his own work, let alone a group of five or six others. Second, if you can find someone talented enough to be a chief programmer, chances are that you can't pay him enough. Someone with the talents we described certainly deserves at least $100,000 per year, but what are the chances that your company would be willing to pay that kind of money?

Third, even if you could pay such a salary, there's a good chance that a chief programmer wouldn't want to work for your organization. After all, you've probably got only one or two "ordinary" mainframe computers, and you're probably developing the same old humdrum accounts receivable and payroll applications that the chief programmer has already implemented dozens of times. He's looking for a company with a football field full of advanced hardware and challenging projects that can really make use of his skills.

The fourth and final reason is that even if you did find a chief programmer, what would you do with all of the average people in your organization? Fire them? Hardly! That's considered antisocial in most large organizations today; you're stuck with them as long as they want to work for you.

So even though the chief programmer team concept has some appeal, there's not much chance that you'll actually be using it. It *is* possible, of course, that you may decide to give your best programmer—someone who is 10 percent better than the average rather than 10 *times* better—a new title of chief programmer. And you may decide that some of the other roles described here would be useful on an EDP project. But don't delude yourself into thinking that you'll get the kind of results that people ascribed to true chief programmer teams.

There's another kind of programming team that is often discussed in the literature: the egoless team. The expression "egoless programming" was coined years ago by Gerald Weinberg (1971). Although it is rarely practiced in American EDP organizations, the concept has a certain fascination to it. Basically, the egoless team is one without a boss (in the conventional sense of the word). The egoless team is composed entirely of peers, a group that manages to run its own day-to-day affairs without direct supervision from anyone else.

As you can imagine, there are many advantages and disadvantages to a project structure of this sort. But because installation of such a team is unlikely to be feasible in most EDP organizations, we won't pursue a discussion of the egoless team at length here. However, it is important to note one aspect of the egoless team that is relevant to most business environments: the willingness of its team members to review and critique each other's work. This extremely valuable notion of peer group review is manifested in a *walkthrough,* which we discuss next.

2.7 STRUCTURED WALKTHROUGHS

The final technique that we will discuss in this chapter is the structured walkthrough. Quite simply, a structured walkthrough is an organized procedure for a group of peers (systems analysts, computer programmers, and so on) to review a technical product for correctness and quality.

Walkthroughs can take place at a number of stages in a typical systems development project, as we will see in later chapters. In much of the recent computer literature, the emphasis has been on *code* walkthroughs (primarily because coding has often been treated as a "private" activity in many EDP organizations, so the idea of "opening up" the coding activity to public inspection has caused great controversy), but the walkthrough concept is just as applicable to reviews of systems designs and specifications. Indeed, walkthroughs at the early stages of a systems development project are crucial: It is devastating to a programmer's morale to be told in a code walkthrough that the code is based on a flawed design or an incorrect specification.

Many managers confuse the concept of a walkthrough with the more conventional "review." There is a simple difference between the two: A walkthrough is conducted by members of a team who work together on a day-to-day basis; reviews, by contrast, involve the presentation of a piece of work to an audience that usually includes the producer's manager and various people who may not be directly involved in the project itself.

Consequently, it is fair to say that walkthroughs are usually characterized by an informal, shirt-sleeve environment, participation by people who are committed to the success of the product, and fast turnaround. By contrast, formal reviews are often attended by people who may not be committed to the success of the product, may not even be familiar with the product, and are characterized by slow turnaround. If errors are found during a walkthrough, another one can usually be scheduled within a day or even a few hours. By contrast, it can take weeks or even months to gather all of the participants together for a second formal review.

Many organizations combine the best of both worlds by having walkthroughs *and* reviews. The walkthrough approach permits fast, effective reviews of the technical correctness and accuracy of the specification, the design, and the code for the system; the formal reviews provide an official stamp of approval, providing the opportunity to allow inspection by quality assurance groups, auditors, people from the computer operations department, and others.

I will assume throughout this book that walkthroughs are used to review the outputs of various activities in the structured project life cycle. For more information on walkthroughs, consult Yourdon's *Structured Walkthroughs* (1985) or the work of Weinberg and Freedman (1977).

2.8 SUMMARY

Although the technical discussion in each section of this chapter was rather brief, you should now have at least a passing familiarity with the major structured techniques. My intent was not to teach you how actually to *practice* structured analysis or structured design but rather to demonstrate that the structured techniques are neither magic nor so terribly complex that they are beyond the grasp of someone outside the technical ranks. Indeed, my intent was to show that structured analysis, structured design, and the other techniques are based on a few commonsense principles, supported by enough rigor and formality to ensure that they work in complex situations. If you've grasped the commonsense part, you should have all you need as a manager; among other things, you should be in a position to know whether your systems analysts and systems designers are tying to pull the wool over your eyes!

With this background, we can now move on to the area where managers tend to feel more at home than technicians, the project life cycle. What are the major activities of a project? What are the interactions between the activities? In what sequence do they need to be done? These are the issues addressed in Chapter 3 and the subsequent chapters.

3

The Structured Life Cycle

3.1 INTRODUCTION

With this chapter, we arrive at the main subject of the book, the structured project life cycle. We begin by discussing the basic concepts of a life cycle and address two questions: What is the purpose of the project life cycle? Why do we have one?

Before introducing the structured project life cycle, it is important to examine the classical project life cycle used in many EDP organizations today, primarily to identify its limitations and weaknesses. This examination will be followed by a brief discussion of the so-called *semistructured* project life cycle: a project life cycle that includes some, but not all, of the elements of structured systems development.

Next, we introduce the *structured* project life cycle, presenting an overview to show the major activities and how they fit together. In subsequent chapters we will examine in considerably greater detail each of the major activities: survey, analysis, design, implementation, test generation, and quality assurance.

Finally, we will examine briefly the *prototyping* life cycle popularized by Bernard Boar, James Martin, and several vendors of fourth-generation programming languages.

We will also explore the concept of *iterative* or *top-down* development in greater detail than in Chapter 2. In particular, we will introduce the notion of *radical* top-down development and *conservative* top-down development. Depending on the nature of your project, there may be valid reasons for adopting one approach rather than the other. It is even possible that your situation calls for a combination of the two.

3.2 THE CONCEPT OF A PROJECT LIFE CYCLE

As a consultant, I have visited hundreds of EDP organizations during the past 20 years, organizations ranging in size from few to thousands of people, with anywhere from 1 to 1,000 projects simultaneously under way. As you might expect, the smaller organizations tend to be relatively informal: EDP projects are begun as the result of verbal discussion between the user and the project manager (who may also be the systems analyst, programmer, computer operator, and janitor!), and the project proceeds from systems analysis through design and implementation without much fuss.

In the larger organizations, however, things are done on a much more formal basis. The various communications between users, management, and the project team tend to be documented in writing, and everyone understands that the project will go through several phases before it is complete. Even so, I have often been surprised by the major differences between the way two project managers in the same organization will conduct their respective projects. Indeed, it is often left to the discretion of the individual project manager to determine what phases and activities his project will consist of and how these phases will be conducted.[1]

Recently, though, the approach taken to systems development has begun to change. More and more large *and* small organizations are adopting a single, uniform project life cycle, sometimes known as a project plan systems development methodology, or simply "the way we do things around here." Usually contained in a notebook as ponderous as the standards manual that sits (unread) on every programmer's desk, the documented project life cycle provides a common way for everyone in the EDP organization to go about the business of developing a computer system.

The approach may be home-grown, or alternatively, the EDP organization may decide to purchase a project management package and then tailor

[1]This sounds as though anarchy prevails in most EDP organizations and nobody has tried to impose any control. However, there are two common situations that lead to this individualistic approach even in the most exemplary organization: (a) the highly decentralized organization, where every department has its own EDP group with its own local standards, and (b) the period of several years immediately after the last "official project life cycle" was deemed a failure and thrown out.

it to company needs.[2] It seems apparent that aside from providing employment for the people who create project life cycle manuals (and for those who write textbooks about them!), the project methodology is desirable. What, then, is the purpose of having a project life cycle? There are three primary objectives:

1. To define the activities to be carried out in an EDP project
2. To introduce consistency among many EDP projects in the same organization
3. To provide checkpoints for management control and checkpoints for "go/no-go" decisions

The first objective is particularly important in a large organization in which new people are constantly entering the ranks of project management. The fledgling project manager may overlook or underestimate the significance of important project phases if he follows only his intuition. Indeed, it can happen that junior programmers and systems analysts may not understand where and how their efforts fit into the overall project unless they are given a proper description of *all* phases of the project.

The second objective is also important in a large organization. For higher levels of management, it can be extremely disconcerting to try to supervise 100 different projects, each of which is being carried out in a different way. For example, if project A defines the systems analysis activity differently than project B does, and project B doesn't include a design phase, how is the second-level or third-level manager to know which project is in trouble and which is proceeding on schedule?

The third objective of a standard project life cycle pertains to management's need to control a project. On trivial projects, the sole checkpoint is likely to be the end of the project: Was it finished on time and within the specified budget? But for larger projects, management should have a number of intermediate checkpoints that provide it with opportunities to determine whether the project is behind schedule and whether additional resources need to be procured. In addition, an intelligent user will also want checkpoints at several stages in the project so that he can determine whether he wants to continue funding it![3]

[2]There are at least half a dozen such packages on the market, costing anywhere from $10,000 to $100,000 or more. Some of the better-known packages are Specrum, SDM-70, and Method/1. For a variety of reasons, I won't comment in this book on any specific project management package; I will only suggest that you keep the concepts presented in this book in mind if you select a vendor-supplied package. You will find, for example, that many of the vendor-supplied packages have incorporated structured techniques in their material, but some have not.

[3]In fact, the politics of most EDP projects are such that there is only one checkpoint at which the user has an obvious, clean way of backing out: at the end of the survey, or feasibility

Having reported all this, let me remind you that the project life cycle is definitely not going to run your project for you. It will not relieve you of the difficult responsibility of making decisions, weighing alternatives, fighting political battles, negotiating with recalcitrant users, boosting the morale of dejected programmers, or any of the other trials and tribulations that face a project manager. If you're going to be the project manager, you still have to *manage,* in every sense of the word. The only help that the project life cycle can provide is that it can *organize* your activities, making it more likely that you'll address the right problems at the right time.

3.3 THE CLASSICAL PROJECT LIFE CYCLE

Before we examine the structured project life cycle, I'd like to make a distinction. If you have worked in the data processing field for more than a few years, there's a good chance that much of this information about project methodologies is familiar to you. You may already have a project life cycle in your organization, and you probably have a number of opinions about it. But I suspect that the kind of project life cycle that your organization is using right now differs from the one to which we'll be devoting most of our attention in this book.

The classical or conventional project life cycle is shown in Figure 3.1. Most of the phases, or activities, should be fairly familiar to you, and I won't bother discussing them in detail here. Certainly every project, whether structured or not, goes through some kind of systems analysis, design, and implementation, even if it's not done in exactly the way I show it here. The project life cycle used in your organization, for example, might differ from the one shown in Figure 3.1 in one or all of the following ways: The survey phase and the analysis phase may be lumped together into a single phase (this is especially common in organizations in which anything the user wants is deemed at the outset to be feasible). There may not be a phase called "hardware study" if it can be taken for granted that any new system can be implemented on an existing computer without causing any major operational impact. The preliminary design and detail design phases may be lumped together in a single phase simply called design, or several of the testing phases may be grouped into a single phase; indeed, they may even be included with "coding." Thus your project life cycle may have 5 phases or 7 phases or 12 phases, but it is still likely to be of the classical variety.

What really characterizes a project life cycle as being classical? Two features stand out: a strong tendency toward bottom-up implementation of

---

study, phase. In theory, though, the user should have the opportunity to cancel an EDP project at the end of any phase if he thinks he is wasting his money.

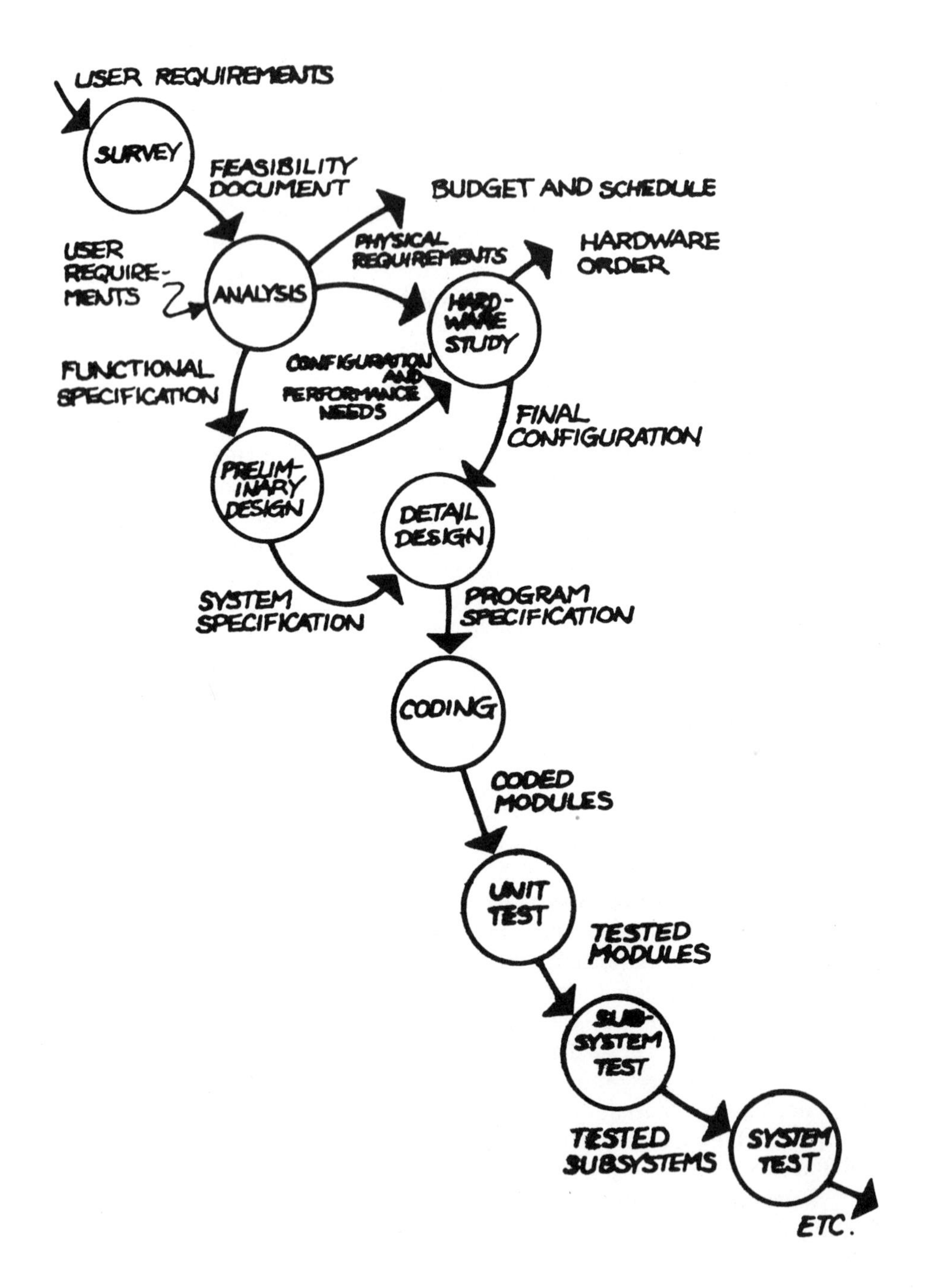

Figure 3.1 The classical project life cycle.

the system and an insistence on linear, sequential progression from one phase to the next.

3.3.1 Bottom-up Implementation

The use of bottom-up implementation is, in my opinion, one of the major weaknesses in the classical project life cycle. As you can see from Figure 3.1, the project manager is expected to carry out all of his module testing first, then subsystem testing, and finally system testing. I'm not quite sure where this approach originally came from, but I wouldn't be surprised if it was borrowed from assembly-line manufacturing industries. The bottom-up implementation approach is a good one for assembling automobiles on an assembly line—*but only after the prototype model has been thoroughly debugged!* Unfortunately, most of us in the computer field are still producing one-of-a-kind systems, for which the bottom-up approach has a number of serious difficulties:

- Nothing is done until it's *all* done. If the project gets behind schedule and the deadline falls right in the middle of system testing, there will be nothing to show the user except an enormous pile of program listings, which, taken in their entirety, do nothing of any value for the user!
- The most trivial bugs are found at the beginning of the testing period, and the most serious bugs are found last. This is almost a tautology: Module testing uncovers relatively simple logic errors inside individual modules; system testing, by contrast, uncovers major interface errors between subsystems. The point is that major interface errors are *not* what you want to find at the end of a development project; they can mean recoding of large numbers of modules and can have a devastating impact on the schedule, right at a time when everyone is likely to be tired and cranky from having worked so hard for so many months.
- Debugging tends to be extremely difficult during the final stages of system testing. Note that we distinguish here between *testing* and *debugging*. Debugging is the black art of discovering *where* a bug is located (and the subsequent determination of how to fix the bug), after the process of testing has determined that there *is* a bug. When a bug is discovered during the system-testing phase of a bottom-up project, it's often extremely difficult to tell which module contains the bug—it could be in any one of the hundreds (or thousands) of modules that have been combined for the first time. Tracking down the bug is often like looking for a needle in a haystack.
- The requirement for test time usually rises exponentially during the final stages of testing. More specifically, the project manager often finds that

he needs large chunks of computer time for system testing—perhaps 12 hours of uninterrupted computer time per day. Since such a large amount of computer time is often difficult to obtain,[4] the project often falls seriously behind schedule.

3.3.2 Sequential Progression

The second major weakness with the classical project life cycle is its insistence that the phases proceed sequentially from one to the next. There is a natural, human tendency to want this to be so: We want to be able to say that we have *finished* the analysis phase and that we'll never have to worry about that phase again. Indeed, many organizations formalize this notion with a ritual known as "freezing the specification" or "freezing the design document."

The only problem with this desire for orderly progression is that it's completely unrealistic! In particular, the sequential approach doesn't allow for real-world phenomena having to do with personnel, company politics, or economics. For example, the person doing the work—such as the systems analyst or designer—may have made a mistake and produced a flawed product. Indeed, as human beings, we rarely do a complex job right the first time, but we are very good at making repeated improvements to an imperfect job. Or the person reviewing the work—in particular, the user who reviews the work of the systems analyst—may have made a mistake. Or perhaps the person carrying out the work associated with each phase may not have enough time to finish but may be unwilling to admit that fact. This is a polite way of saying that on most complex projects, systems analysis and design (and system testing, too) finish when someone decides that you have run out of time, not when you want them to finish!

Other problems are commonly associated with the sequential, classical project life cycle. During the several months (or years) that it takes to develop the system, the user may change his mind about what he wants the system to do. During the period that it takes to develop the system, certain aspects of the user's environment, such as the economy, the competition, or the government regulations that affect the user's activities, may change.

An additional characteristic of the classical project life cycle is that it relies on outdated techniques; that is, it tends to make no use of structured design, structured programming, walkthroughs, or any of the other modern development techniques. But because the classical project life cycle *ignores* the existence of these techniques, there is nothing to prevent the project manager

[4]I'm convinced that yet another of the Murphy-type laws applies in this regard: The larger and more critical the project, the more likely it is that its deadline will coincide with end-of-year processing and other organizational crises that gobble up all available computer time!

from using them. Unfortunately, many programmers, systems analysts, and first-level project leaders feel that the project life cycle is a statement of policy by top-level management—and if management dosn't say anything about the use of structured programming, then they, as mere project members and leaders, are not obliged to use nonclassical approaches.

3.4 THE SEMISTRUCTURED LIFE CYCLE

My comments make it seem as if most EDP organizations are still living in the Dark Ages. Indeed, that picture is somewhat unfair; not *every* organization uses the classical project life cycle. Particularly during the 1980s there has been a growing recognition that techniques like structured design, structured programming, and top-down implementation should be officially recognized in the project life cycle. This recognition has led to the semistructured project life cycle shown in Figure 3.2.

As you can see, the life cycle in the figure shows two obvious features not shown in the classical approach:

- The bottom-up sequence of coding, module testing, and system testing is replaced by top-down implementation, as discussed in Chapter 2. There is also a strong indication that structured programming is to be used as the method of actually coding the system.
- Classical design is replaced by structured design, as discussed in Chapter 2.

Aside from these obvious differences, there are some subtle points about this modified life cycle. Consider, for example, that top-down implementation, as described in Chapter 2, means that some coding and testing are taking place in parallel. That certainly represents a major departure from the sequential phases that we saw in the classifcal life cycle! In particular, it can mean *feedback* between the activity of coding and that of testing and debugging. When testing the top-level skeleton version of his system, the programmer may be heard to mutter to himself, "Jeez, I had no idea that the double-precision frammis instruction worked *that* way!" Naturally, you can be sure that his subsequent use of the double-precision frammis instruction will be quite different.

Perhaps more important, the use of top-down implementation tempts the implementers (and the systems analysts, if they haven't abandoned the project by this time) to talk to the users *after* the specifications have been ceremoniously frozen. Thus it is possible that the user will point out errors or misunderstandings in the specifications; he may even express a desire to *change* the specifications—and if the conversation takes place directly between

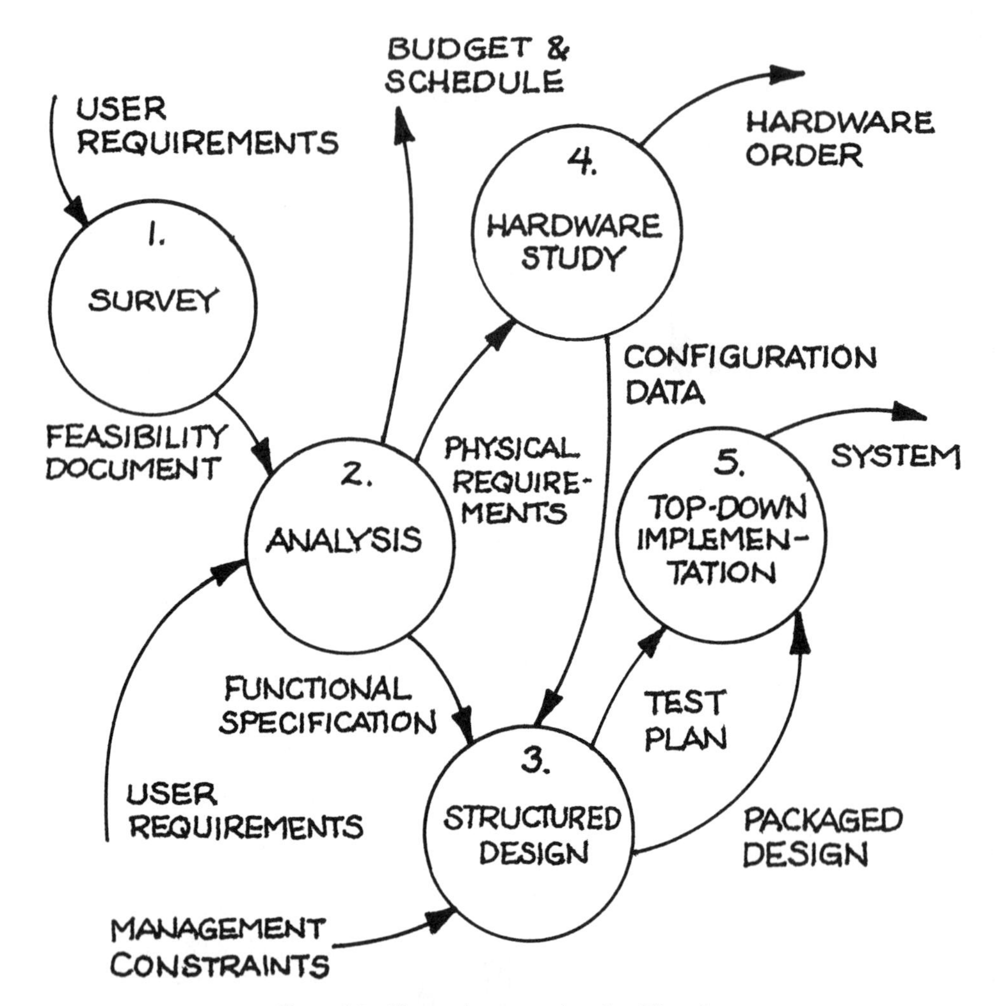

Figure 3.2 The semi-structured project life cycle.

the user and the implementer, a change may actually be effected before EDP project management knows what is happening. In short, top-down implementation often provides feedback between the implementation process and the analysis process—although this feedback exchange is not specifically shown in Figure 3.2 and although the user and the EDP project manager might well deny that it is taking place!

There is one final subtle point about the semistructured life cycle. A significant part of the work that takes place under the heading of "structured design" is actually an effort to fix up bad specifications. You can see this by looking at Figure 3.3, a diagram depicting the detailed of structured design.

In Figure 3.3, activity 3.1 (labeled CODIFY FUNCTIONAL SPECIFICATION)

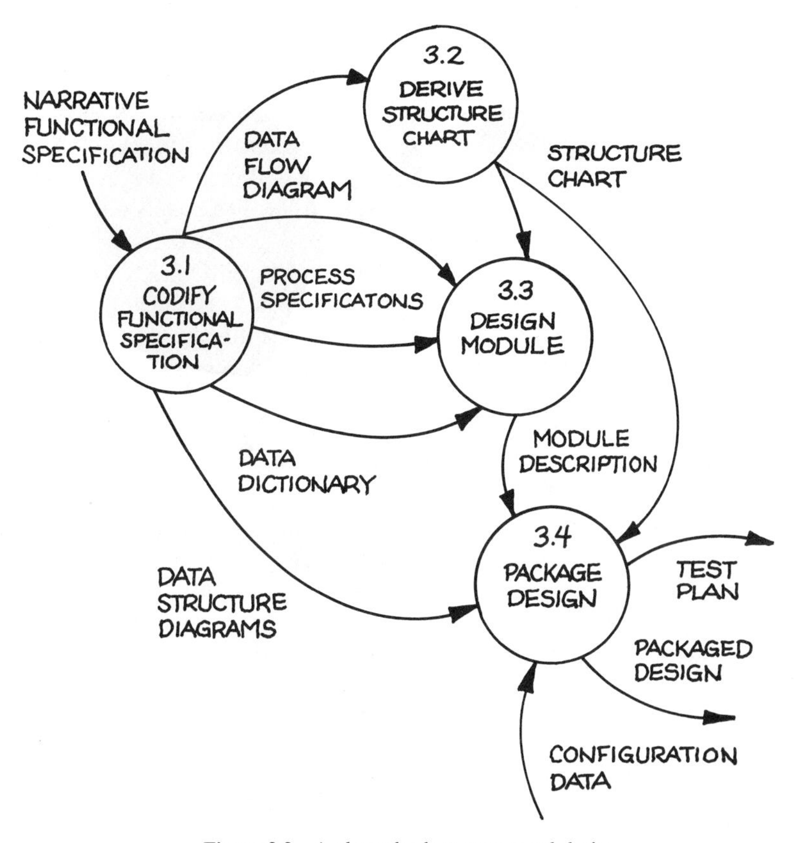

Figure 3.3 A closer look at structured design.

represents a task that designers have long had to do: translate a monolithic, ambiguous, redundant, narrative document into a useful, nonprocedural model to serve as the basis for deriving the module hierarchy. In other words, people practicing structured design have traditionally assumed that they would be given a classical specification; consequently, their first job, as they see it, is to transform that specification into a package of data flow diagrams, data dictionary entries, entity relationship diagrams, and process specifications. This is a more diffficult job than you might imagine: Historically, it has been a task carried out in a vacuum. Designers generally had little contact with the systems analyst who wrote the long narrative specification, and he certainly had *no* contact with the user.

Obviously, such a situation is ripe for change. Introducing structured analysis into the picture, as well as expanding on the idea of feedback between one part of the project and another, creates an entirely different kind of project life cycle. This is the structured project life cycle, which we will discuss in the next section.

3.5 THE STRUCTURED PROJECT LIFE CYCLE

Now that we have seen the classical project life cycle and the semistructured project life cycle, we are ready to examine the structured life cycle, our topic for the remainder of this book. This life cycle is shown in Figure 3.4.

At this point, let us look briefly at the project life cycle's nine activities and three terminators, as shown in Figure 3.4. Let's begin by defining the outside entities on the diagram—the rectangular boxes labeled USERS, MANAGEMENT, and OPERATIONS. These are individuals or groups of individuals who provide input to the project team and who are the ultimate recipients of the system. Although the purpose of this book is to concentrate on the task of building an EDP system, we must clarify what we mean by the words *user, management,* and *operations.*

What do we mean, for example, when we say "user"? Many people suggest that "customer" or "client" might be a more apt name, since this is the person who ultimately pays for the system (and thus, indirectly, pays the salaries of the people who build the system). But by whatever name we choose to call him, there is one crucial property unique to the user: Only he can accept the system when it is finished. Only he can decide whether it is suitable and whether it will be integrated into his existing business operation. As we will see in sub-

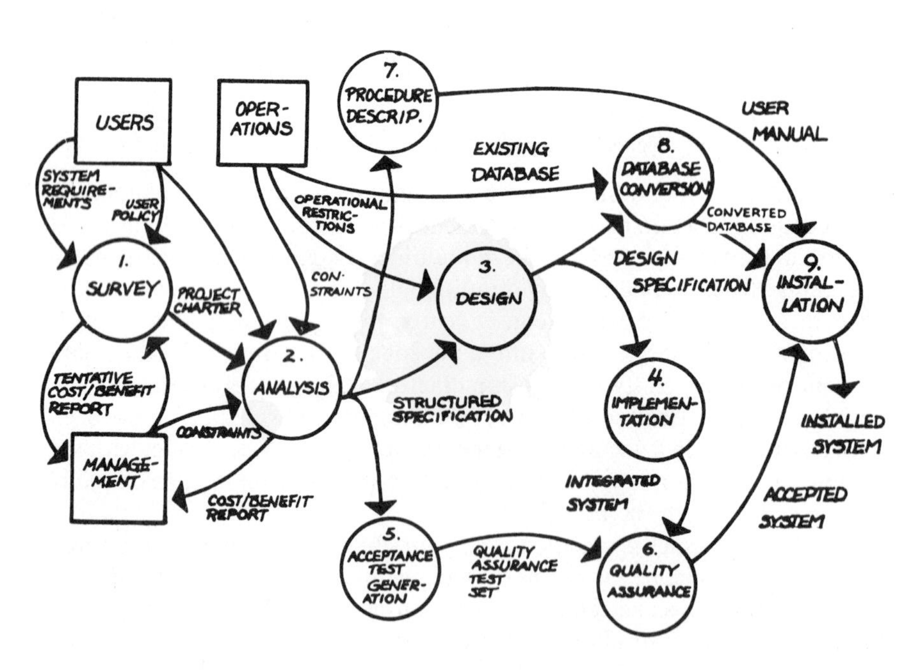

Figure 3.4 The structured project life cycle.

sequent chapters, this has a significant impact on the business of acceptance testing (activities 5 and 6).

One other distinction should be made about users: In most cases, there are different *types* of users who come into contact with the EDP system during its formulation, its development, and its actual operational use. It's useful to identify three different classes of users who interact with the project team that builds an EDP system: the *strategic* user, the *tactical* user, and the *operational* user.

The *strategic* user is primarily concerned with the long-term profitability or return on investment that will be generated by the proposed new system. This person is often head of an operational division—for example, the manager of accounting or the vice-president of manufacturing—and is primarily concerned with the survey phase (activity 1) and the early stages of systems analysis (activity 2).

The *tactical* user is the person who supervises the day-to-day operations of some piece of the business and whose performance is usually measured against a budget. This person usually has a good overview of the day-to-day functioning of his part of the business, though he may not be intimately familiar with the lowest-level procedural details. The tactical user usually supervises at least one administrative or clerical worker and is concerned that the development of a new EDP system not interfere with his subordinates' work; hence the systems analyst will often hear the tactical user exclaim, "Don't talk to my people—they're far too busy! I'll tell you everything you need to know about their jobs!"

The *operational* user is the clerical or administrative person found so often in modern organizations. This is the person who will have the most contact with the EDP system once it has become operational. For on-line systems, this is the person for whom such things as man-machine dialogues and error messages are most important.

Just as the term *user* covers a broad spectrum, so too does the term *management* as we have used it to represent an outside entity in Figure 3.4. In some cases, user and manager are one and the same; indeed, it may well turn out that management is what we just described as the strategic user. In other cases the term might represent a higher authority—the executive committee, a steering committee that reviews all EDP projects, or even the board of directors. In the context of this book, we are interested in identifying a person (or group) who approves the funding of the project and provides *organizational* constraints.[5]

The third terminator shown in Figure 3.4 is simply labeled "operations."

[5]An example of such organizational constraints might be the following message to the project team: "We don't care what the users want; we have a company policy that *all* projects will be implemented on Blatzco computers."

Again, this can mean several things. In the case of a small, stand-alone minicomputer or microcomputer system, the user may also be the *operator*. In most cases, though, it is meaningful to distinguish between the two and to identify the person or group responsible for the day-to-day operation of the computer hardware and software that will comprise the system being built. In the vast majority of business-oriented EDP projects, the new system will operate, along with dozens of other systems, on computer hardware controlled by a centralized group (even though the hardware itself may be decentralized). Hence our model of the system for building systems should show this interaction, and the project team should be prepared for comments like "We don't care what the users want, and we don't care what management says; all we care about is that whatever system you develop doesn't use more than 4 megabytes of memory."

These three entities—users, management, and operations—interact with the nine activities that we have shown in Figure 3.4. Let's summarize each of these activities:

1. *Survey.* This activity is also known as the feasibility study or initial business study. Typically, it begins when a user requests that one or more portions of his business be automated. The major purpose of the survey activity is to identify current deficiencies in the user's environment; to establish new goals; to detemine whether it is feasible to automate the business and, if so, to suggest some acceptable scenarios; and finally, to prepare a project charger that will be used to guide the remainder of the project.
2. *Analysis.* The primary purpose of the analysis activity is to transform its two major inputs, user policy and the project charger, into a structured specification. This involves modeling the user's environment with data flow diagrams and the other tools presented in Chapter 2.
3. *Design.* The activity of design is concerned with allocating portions of the specification (otherwise known as the "essential model") to appropriate processors (CPUs and/or humans) and to appropriate tasks (or jobs, or partitions, etc.) within each processor. Within each task, the design activity is concerned with the development of an appropriate hierarchy of program modules and interfaces between those modules to implement the specification created in activity 2.
4. *Implementation.* This activity includes both coding and the integration of modules into a progressively more complete skeleton of the ultimate system. Thus activity 4 includes both structured programming and top-down implementation.
5. *Acceptance test generation.* The structured specification should contain all of the information necessary to define an acceptable system, from

the user's point of view. Thus once the specification has been generated, work can commence on the activity of generating a set of acceptance test cases from the structured specification.

6. *Quality assurance.* Quality assurance is also known as final testing or acceptance testing. This activity requires, as its input, acceptance test data generated in activity 5 and an integrated system produced by activity 4.
7. *Procedure description.* Throughout this book, we concern ourselves with the development of an *entire* system—not just the automated portion but also the portion to be carried out by people. Thus one of the important activities to be performed is the generation of a formal description of the portions of the new system that will be manual, as well as a description of how the users will actually interact with the automated portion of the new system. The output of activity 7 is a user's manual.
8. *Database conversion.* In some projects that I have been associated with, database conversion involved more work (and more strategic planning) than the development of computer programs for the new system; in other cases there might not have been any existing database to convert. In the general case, this activity requires, as input, the user's current database, as well as the design specification produced by activity 3.
9. *Installation.* The final activity, of course, is installation; its inputs are the user's manual produced by activity 7, the converted database produced by activity 8, and the accepted system produced by activity 6. In some cases, however, installation may simply mean an overnight cutover to the new system, with no excitement or fanfare; in other cases installation may be a gradual process, as one user group (or distributed location) after another receives user's manuals, hardware, and training in the use of the new system.

It's important that you view Figure 3.4 for what it is: a *data flow diagram.* It is *not* a flowchart; there is no implication that all of activity N must finish before activity N + 1 commences. On the contrary, the network of data flows connecting activities strongly implies that several activities may be going on in parallel. It is because of this nonsequential aspect that I use the word *activity* in the structured project life cycle, rather than the more conventional word *phase. Phase* has traditionally referred to a particular period of time in a project when one, and only one, activity was going on.

Something else must be emphasized about the use of a data flow diagram to depict the project life cycle: A data flow diagram, such as the one in Figure 3.4, does not explicitly show feedback, nor does it show control. Virtually every one of the activities in Figure 3.4 can, and usually does, produce information that can provide suitable modifications to one or more of the preceding activities. Thus the activity of design can produce information that may revise some of the cost/benefit decisions that take place in the analysis activity; indeed, knowledge gained in the design activity may even require revising earlier decisions about the basic feasibility of the project.

In extreme cases, certain events taking place in any activity could cause the entire project to terminate abruptly. The input of management is shown only for the analysis activity because analysis is the only activity that requires *data* from management; it is assumed, however, that management exerts *control* over *all* the activities.

In summary, then, the diagram only tells you the input(s) required by each activity and the output(s) produced. The sequence of activities can be implied only to the extent that the presence or absence of data makes it possible for an activity to commence.

Clearly, Figure 3.4 does not provide you with enough information actually to manage a project. In fact, most of the activities have half a dozen or more subactivities, and there is much more to say about each of them. Chapter 4 explores each project management activity in greater detail.

3.6 RADICAL VERSUS CONSERVATIVE TOP-DOWN IMPLEMENTATION

In the preceding section I pointed out that the structured project life cycle allows more than one activity to take place at one time. In the most extreme situation, *all* of the activities in the structured project life cycle could be taking place simultaneously. At the other extreme, the project manager could decide to adopt the sequential approach, finishing all of one activity before commencing the next.

I've found that it's useful to have some terminology to help talk about these extremes, as well as about compromises between the two extremes. A *radical* approach to the structured project life cycle is one in which activities 1 through 9 take place in parallel from the very beginning of the project: Coding begins on the first day of the project, and the survey and analysis continue until the last day of the project. By contrast, in a *conservative* approach to the structured project life cycle, all of activity N is completed before activity N + 1 begins.

Obviously, no manager in his right mind would adopt either of these two extremes. The key is to recognize that the radical and conservative extremes are the two endpoints in a range of choices. This is illustrated in Figure 3.5.

Keep in mind when you begin your next project that there are an infinite number of choices between the radical and conservative extremes. You might decide to finish 75 percent of the survey activity, followed by completion of 75 percent of analysis and then 75 percent of design, in order to produce a

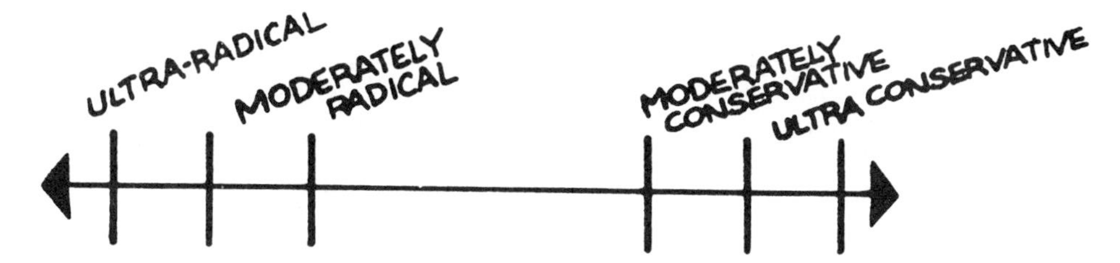

Figure 3.5 Radical to conservative range.

reasonably complete skeleton version of a system whose details could then be refined by a second pass through the entire project life cycle. Or you might decide to finish all of the survey and analysis activities, followed by completion of 50 percent of design and 50 percent of implementation. The possibilities are truly endless. How should you, as a project manager, decide whether to adopt a radical or conservative approach on your next project? Basically, there is no right answer. You must base your decision on the following factors:

- How fickle is the user?
- What pressure are you under to produce immediate, tangible results?
- What pressure are you under to produce an accurate schedule, budget, and estimate of manpower and other resources?
- What are the dangers of making a major technical blunder?

As you can appreciate, not one of these questions has a straight black-or-white answer. For example, you can't ask the user of the system, in casual conversation, "By the way, how fickle are you feeling today?" But you should be able to assess the situation by observation, especially if you're a veteran project manager who has dealt with many users and upper-level managers before.

If you judge that you're dealing with a fickle user, one whose personality is such that he delays final decisions until he sees how the system is going to work, you would probably opt for a more radical approach.[6] The same is true if you're dealing with an inexperienced user, one who has had very few systems built for him. Why spend years developing an absolutely perfect set of specifications only to discover that the user didn't understand the significance of the specifications?

If, however, you're dealing with a veteran user who is absolutely sure of what he wants, and if he works in a business area that is stable and unlikely to change radically on a month-to-month basis, you can afford to take a more conservative approach. Of course, there are a lot of in-between situations:

[6]Alternatively, you might consider the prototyping life cycle discussed in Section 3.7.

The user may be sure of *some* of the business functions to be performed but somewhat unsure of the kinds of reports and management information he would like the system to provide. Or if he is familiar with batch computer systems, he may be unsure of the impact that an on-line system will have on his business.

Besides fickleness, there is a second factor to consider: the pressure to produce immediate, tangible results. If, due to politics or other external pressures, you simply *must* get a system up and running by a specific date, a somewhat radical approach is warranted. You still run the risk that your system will be only 90 percent complete when the deadline arrives, but at least it will be a *working* 90 percent complete skeleton that can be demonstrated and perhaps even put into production. That's generally better than having finished all of the systems analysis, all of the design, and all of the coding but none of the testing.

Of course, all projects are under some pressure for tangible results—it's simply a question of degree. And it's an issue that can be rather dynamic. A project that begins in a low-key fashion with a comfortable schedule can suddenly become a high priority, and the deadline may be advanced six months to a year. One of the advantages of doing the analysis, design, coding, and implementation from the top down is that one can stop an activity at any point and leave the remaining details for subsequent consideration; meanwhile, the top-level analysis that has been completed can be used to begin the top-level design—and so forth.

Yet another factor in project management is the ever-present requirement, in most large organizations, to produce schedules, estimates, budgets, and the like. In some organizations this tends to be done in a fairly informal fashion, typically because the projects are relatively small and because management feels that any errors in estimating will have an insignificant impact on the whole organization. In such cases one can adopt a radical approach, even though any attempts at estimating will have to be "gut-level" guesses. By contrast, most large projects require relatively detailed estimates of manpower, computer resources, and so on, and this can be done only after a fairly detailed survey, analysis, and design have been completed. In other words, the more detailed and accurate your estimates have to be, the more likely you are to follow a conservative approach.

Finally, you should consider the danger of making a major technical blunder. For example, suppose that all of your past experience as a project manager has been with a small batch-oriented System/36 computer system, and now, all of a sudden, you find yourself developing an on-line, real-time, multiprocessing distributed database management information system that will process 2 million transactions a day from 5000 terminals scattered around the world. In such a situation one of the dangers of a radical approach is discovering a major design flaw after a large portion of the top-level skeleton system has been implemented. You may discover, for example, that in order for your whiz-bang system to work, a low-level module has to do its job in 19 microseconds, but your programmers suddenly tell you that there is no way on earth to code the module that efficiently—not in COBOL, not in C, not in assembly language, not even in microcode. So you must be alert to the fact that following the radical approach requires you to pick a "top" to your system relatively early in the game, and there is always the danger of discovering, down toward the bottom, that you picked the wrong top!

However, consider another scenario: You've decided to build an EDP system with new hardware, a new operating system, a new database management system (produced by someone other than the hardware vendor), and a new telecommunications package (produced by yet another vendor). All of the vendors have impressive, glossy manuals describing their products, but the vendors have never interfaced their respective hardware and software products. Who knows if they will work together at all? Who knows if the throughput promised by one vendor will be destroyed by the system resources used by one of the other vendors? Certainly, in a case like this, the project manager might elect a radical approach, so that a skeleton version of the system could be used to explore possible interface problems between the vendors' components.

If you're building a familiar kind of system, such as your ninety-ninth payroll system, you probably have a very good idea of how realistic your goals are. You probably remember from your last project what sort of modules you're going to need at the detailed level, and you probably remember very clearly what the top-level structure looked like. In such a case you may be willing to accept the risks of making a mistake because of the other benefits that the radical approach will give you.

In summary, the radical approach is most suitable for thinly disguised research and development efforts. It is good in environments in which something *must* be working on a specific date and in situations where the user's perception of what he wants the system to do is subject to change. The conservative approach tends to be used on larger projects in which massive amounts of money are being spent and for which careful analysis and design are required to prevent subsequent disasters. However, every project is different and requires its own special blend of radical and conservative top-down implementation. To deal with the individual nature of any project, you should be prepared to modify your approach in midstream, if necessary.

3.7 THE PROTOTYPING LIFE CYCLE

A variation on the top-down approach has become popular in the past few years. It is generally known as the *prototyping* approach and has been popularized by Bernard Boar, James Martin, and others. As Boar describes it in *Application Prototyping* (1984):

> An alternative approach to requirements definition is to capture an initial set of needs and to implement quickly those needs with the stated intent of iteratively expanding and refining them as mutual user/developer understanding of the system grows. Definition of the system occurs through gradual and evolutionary discovery as opposed to omniscient foresight. . . . This kind of approach is called prototyping. It is also referred to as system modeling or heuristic development. It offers an attractive and workable alternative to prespecification methods to deal better with uncertainty, ambiguity, and fickleness of real-world projects. (p. 5)

In many ways this sounds exactly like the radical top-down approach discussed in Section 3.6. The primary difference is that the structured approach discussed throughout this book presumes that sooner or later a complete *paper model* of the system will be built, that is, a complete set of data flow diagrams, entity relationship diagrams, process specifications, and so on. The model will be completed sooner with a conservative approach and later with a radical approach, but by the end of the project there will be a formal set of documents that should live forever with the system as it undergoes maintenance and revision.

The prototyping approach by contrast, almost always assumes that the model will be a "working model," a collection of computer programs that will simulate some or all of the functions that the user wants. But since those computer programs are intended just as a model, there is also an assumption that *when the modeling is finished, the programs will be thrown away and replaced with "real" programs.* The prototyping approach typically uses one or more of the following kinds of software tools:

- Integrated data dictionary
- Screen generator
- Nonprocedural report writer
- Fourth-generation programming language
- Nonprocedural query language
- Powerful database management facilities

The prototyping life cycle proposed by Boar is shown in Figure 3.6. It begins with a survey activity, similar to that proposed in this book; that is immediately followed by a determination of whether the project is a good candidate for a prototyping approach. Good candidates for a prototyping approach are projects that have the following characteristics:

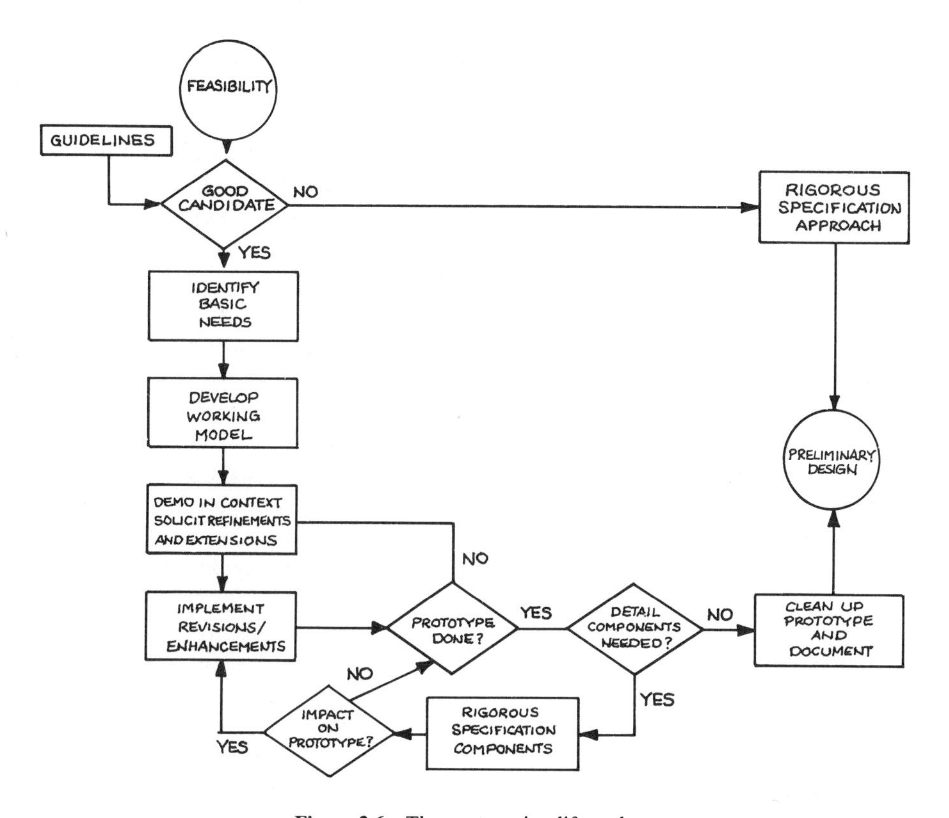

Figure 3.6 The prototyping life cycle.

- The user is unable or unwilling to examine abstract paper models like data flow diagrams.
- The user is unable or unwilling to articulate (''prespecify'') his requirements in any form and can only determine his requirements through a process of trial and error.
- The system is intended to be on-line, with full-screen terminal activities, as opposed to batch edit, update, and report systems. (Almost all prototyping software tools are oriented toward the on-line, database, terminal-oriented approach; there are few vendor-supplied software tools to help build prototypes of batch systems.)
- The system does *not* require specification of large amounts of algorithmic detail, that is, the writing of many, many process specifications to describe the algorithms by which output results are created. Thus decision support, ad hoc retrieval, and record management systems are good candidates for prototyping. Good candidates tend to be the systems in which the user is more concerned about the format and layout of the

CRT data entry screens and error messages than about the underlying computations performed by the system.

It is significant to note that the prototyping life cycle shown in Figure 3.6 concludes by entering the design phase of a "traditional" structured life cycle of the sort described in this book. Specifically, this means that the prototype is *not* intended to be an operational system; it is intended solely as a means of modeling user requirements.

The prototyping approach certainly has merit in a number of situations. In some cases the project manager may want to use the prototyping approach as an alternative to the structured analysis approach described in Chapter 5; however, I think it makes much more sense to use it *in conjunction with* the development of paper models such as data flow diagrams. It is unfortunate that many textbooks and consultants portray prototyping and structured analysis as incompatible, mutually exclusive concepts; indeed, the two concepts can be used together in many cases.

Keep in mind the following points about the prototyping approach:

- The top-down approach described earlier is another form of prototyping, but instead of using vendor-supplied tools such as screen generators and fourth-generation languages, the project team uses the system itself as its own prototype. That is, the various "versions" of a skeleton system provide the "working model" that the user can interact with to get a more realistic feeling of system functions than he might get from a paper model.
- The prototyping life cycle, as described here, involves the development of a working model that is then thrown away and replaced by a "production" system. There is a significant danger that either the user or the development team may try to turn the prototype itself into a production system. With current hardware and software technology, this often turns out to be a disaster because the prototype cannot handle large volumes of transactions efficiently and because it lacks such operational details as error recovery, audit trails, backup/restart facilities, user documentation, and conversion procedures. Obviously, as hardware and software technology continues to improve, this will be less and less of a problem; so-called evolutionary prototyping will probably become practicable by the early 1990s.
- A major problem with some prototyping packages is that they require a completely defined data dictionary before the advantages of the screen generators, report generators, and fourth-generation languages can be employed. But the development of the data dictionary is a tedious, time-consuming activity; this causes great frustration for users and systems

developers who thought that the prototyping approach was a form of magic that would give them "instant" systems. (Note, by the way, that the structured analysis approach also requires the development of a complete data dictionary.)

- If the prototype is indeed thrown away and replaced by the production system, there is a real danger that the project may finish without having a permanent record of user requirements. This is likely to make maintenance and modification increasingly difficult as time goes on (ten years after the system is built, it will be difficult for maintenance programmers to incorporate a change, because nobody, including the "second-generation" users who are now working with the system, will remember what it was supposed to do in the first place). The life cycle presented in this book is based on the idea that the paper models developed in the analysis activity will not only be input to the design activity but will also be retained (and modified, as necessary) during ongoing maintenance throughout the life of the system.

3.8 SUMMARY

The major purpose of this chapter has been to provide an overview of project life cycles in general. If your organization is currently using a formal project life cycle, you should be able to tell whether it falls into the category of classical, semistructured, structured, or prototyping—and you may have already gotten some ideas for modifying your own project life cycle to avoid some of the problems discussed in this chapter.

If you are accustomed to managing projects that are working on only one activity at a time, the discussion of radical top-down implementation and conservative top-down implementation in Section 3.6 may have disturbed you. This was my intent, as the major purpose of that section was to make you think about the *possibility* of overlapping some of the major activities in an EDP project. Obviously, it's more difficult to manage any project that has several activities taking place in parallel, but to some extent that happens in every project. Even if you, as the project manager, decide that your people will concentrate all their efforts on one major activity at a time, there will still be a number of subactivities taking place in parallel. Multiple systems analysts will be interviewing multiple users simultaneously; various pieces of the final product of systems analysis will be in various stages of progress throughout the analysis phase. One of your jobs as a project manager is to exercise sufficient control over those subactivities to ensure that they coordinate smoothly. And in virtually every EDP project that I have observed or participated in this same kind of parallel activity operates at a higher level, too: Regardless of the organization's formal project life cycle, the reality is that many of the major project activities do overlap to some extent. Nevertheless, if you decide to insist on a strictly sequential progression of project activities, the project life cycle presented in this book will still work.

In the next chapter we begin examining the first of the major project activities, the survey, in detail. As we will see, the purpose of this activity is to develop a *project charter,* a document that defines the broad scope within which the project team may work. When you next read about or hear about a major EDP project failure, try to find out whether the project team developed any charter at all; you'll often find that because of the lack of a charter, members of the project team thought that they were engaged in a state-of-the-art research project, while top management thought that the project team was building a mundane production system to generate additional revenues for the organization. The solution: a formal project charter, developed along the lines presented in Chapter 4.

4

Activity 1: Survey

4.1 INTRODUCTION

The first activity in a software development project is the *survey,* or, as it is often called, the *feasibility study.* During the survey, problems and deficiencies in the user's environment are evaluated, with the assumption that they *may* provide the justification for developing new approaches. Figure 4.1 shows the context of the survey.

Note that the diagram in Figure 4.1 shows a small portion of the overall life cycle that was illustrated in Figure 3.4. Note also that the survey activity involves interactions with both the user community and with management.

Figure 4.2 provides a closer look at the survey activity. As you can see, there are four major subactivities: "identify current deficiencies," "establish new system goals," "generate acceptable scenarios," and "prepare project charter."[1]

Perhaps the most difficult thing to define about the survey is its scope: How much of the user's environment should be studied? Scope is of critical importance, because the project charter produced by the survey essentially states the degree of freedom that the systems analysts have in exploring and defining detailed user policy. For example, it is in the survey activity that someone will make the fundamental decision (to be documented in the project

[1] It is important to remember that the *project charter* is the primary deliverable, or product, of the survey. It provides the basis for the detailed systems analysis that follows.

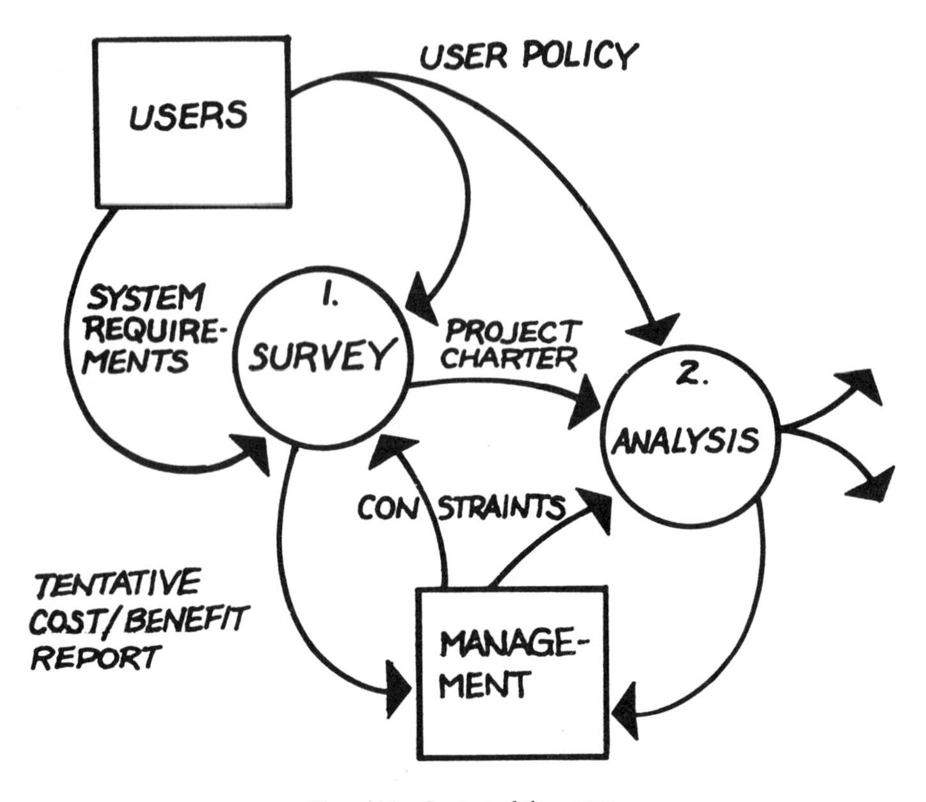

Figure 4.1 Context of the survey.

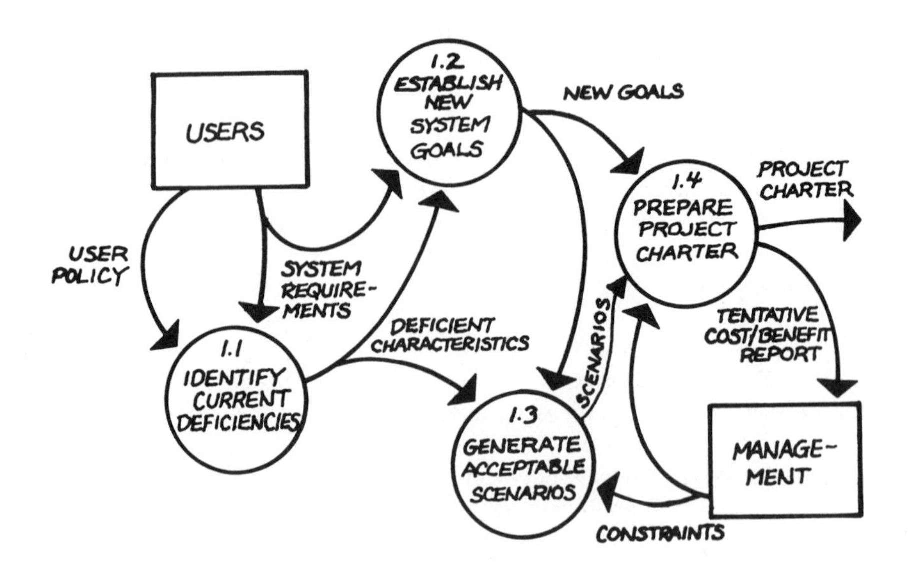

Figure 4.2 Details of the survey activity.

charter) that the purpose of the project is to develop a new accounting system for receivables and payables but that payroll is outside the project's scope.

A veteran systems analyst knows, from painful experience, that a problem (and concomitant request for computerization) in one area of the organization may be nothing more than a symptom of a *real* problem in some other area.[2] Hence he may argue that the scope of the survey should be enlarged to include as much of the organization as possible—and management (or the user who initiated the project) may complain that such an enlarged survey would take too long, cost too much, and so on.

But what the analyst is really trying to do is ensure that a new system, if it is built, will *not* affect some area of the organization outside of his study; areas of the organization affected by a new system should be a proper subset of the area included in the survey. Pictorially, what we want is the situation shown in Figure 4.3(a); what we want to avoid is the situation shown in Figure 4.3(b).

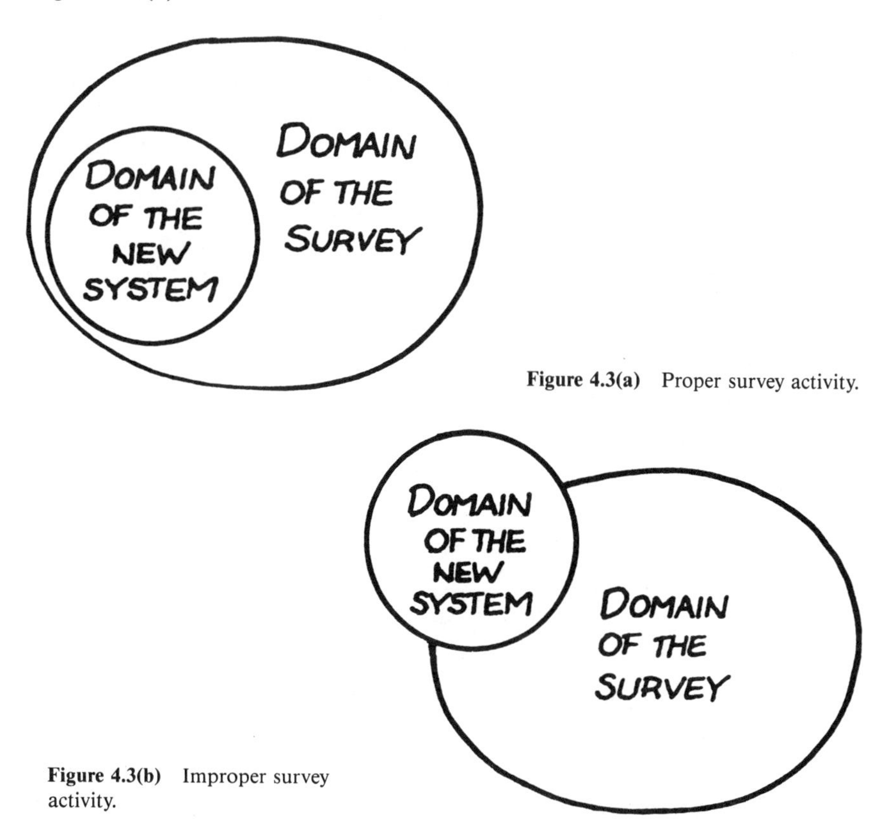

Figure 4.3(a) Proper survey activity.

Figure 4.3(b) Improper survey activity.

[2]Or it may turn out that the automation project reflects the user's fascination with computer technology and his desire to have a personal computer on his desk.

I know of no method that will guarantee users, management, and the people conducting the study that a sufficiently broad scope will be chosen, nor do I know of any way to guarantee that the scope won't be too large. But it certainly is an important issue; it should be discussed openly by all three parties as early as possible, and it should be documented explicitly in the project charter.

4.2 ACTIVITIES OF THE SURVEY

In the subsequent sections of this chapter, we will examine each of the subactivities of the survey in detail.

4.2.1 Subactivity 1.1: Identify Current Deficiencies

There is an assumption about the survey, and indeed about the entire system life cycle, that is so fundamental that we often ignore it: The user would not be asking for a new system unless something was wrong with his present environment. So the first step in the survey activity is to interview the appropriate member(s) of the user community to identify the aspects of the current environment that are inadequate. Figure 4.4 depicts this step.

As you can see from Figure 4.4, this subactivity has two inputs from the user: *user policy* and *system requirements.* We need to define each of these more fully. I will give an abbreviated definition in a data dictionary format first:

- *System requirements:* verbal, unstructured list of required functions and operational characteristics
- *User policy:* verbal or informal written description of current operational characteristics

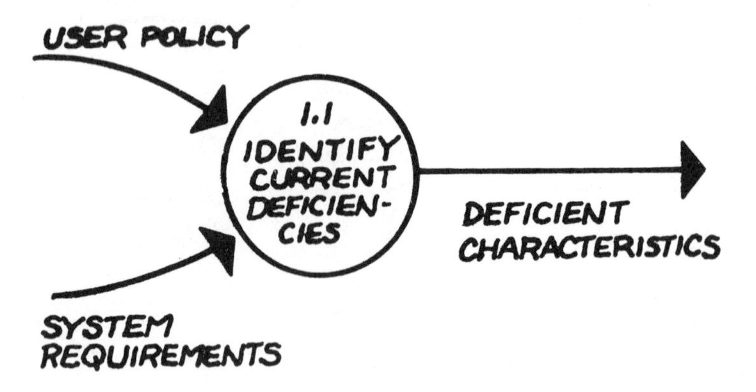

Figure 4.4 Identifying current deficiencies.

- *Deficient characteristics:* missing functions, excessive costs, throughput problems, reliability problems, and operational problems, listed in order of priority.

The system requirements are usually obtained from the user through a series of interviews and dialogues. On a small project, such as one lasting from three to six months, this activity could require one or two interviews with one or two users; on a large project, dozens of interviews with scores of users (in different departments, different locations, and so forth) might be required. The techniques for conducting such interviews are beyond the scope of this book; I recommend that you consult such basic texts as Gildersleeve's *Successful Data Processing Systems Analysis* (1978) for some pointers. You'll find, for example, that many users will complain about *symptoms* rather than *causes* or that they will complain about short-term tactical problems when in reality the long-term strategy of their business is deficient.

Similarly, user policy is obtained through a series of interviews. Depending on the size and formality of the project, it may be appropriate to obtain this information in a written form; indeed, in the extreme case it may take the form of a high-level data flow diagram. In other cases the project may be small enough or the analyst may be sufficiently familiar with the user's environment that the user policy can be communicated in an entirely verbal fashion.

Indeed, the only organized, orderly, *written* document associated with this first subactivity of the survey is the set of deficient characteristics, which is typically written in a narrative style and is relatively short (from one to five pages long).

4.2.2 Subactivity 1.2: Establish New System Goals

The second subactivity of the survey, as shown in Figure 4.5, is the establishment of new goals. This effort takes the form of a concise statement of functional and performance objectives of the new system that is to be built; like the listing of deficient characteristics produced in subactivity 1.1, the statement is usually written in a narrative style and is typically from one to five pages long for medium-sized projects.

Using the data dictionary format, we can define new goals as follows:

- *New goals:* exhaustive list of required functions + throughput/response requirements + cost limitations + reliability/accuracy requirements

Note that the four elements comprising new goals are precisely those likely to be identified as deficient characteristics in the present system. For example,

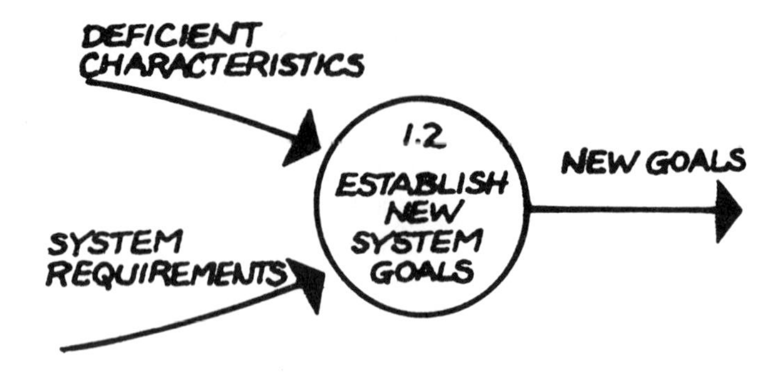

Figure 4.5 Establishing new system goals.

the user may point out that one of the deficiencies in the current system is that it breaks down every 30 minutes and that all work in progress has to be restarted. This presumably would influence him to request that a new system be developed with a mean time between failures (MTBF) of 100 hours, for example.

4.2.3 Subactivity 1.3: Generate Acceptable Scenarios

The purpose of subactivity 1.3 is to provide a brief description of several possible systems that satisfy the user's new goals within the constraints imposed by management; this is shown in Figure 4.6. Although an obvious and important step, the generation of an acceptable system scenario is often done poorly. One particular problem is that the analyst often provides only one scenario for the user's consideration.

We will discuss why the one-scenario presentation is a problem later in this section, but let's first examine inputs and outputs to the subactivity of generating acceptable scenarios. One of the inputs is new goals, which we've

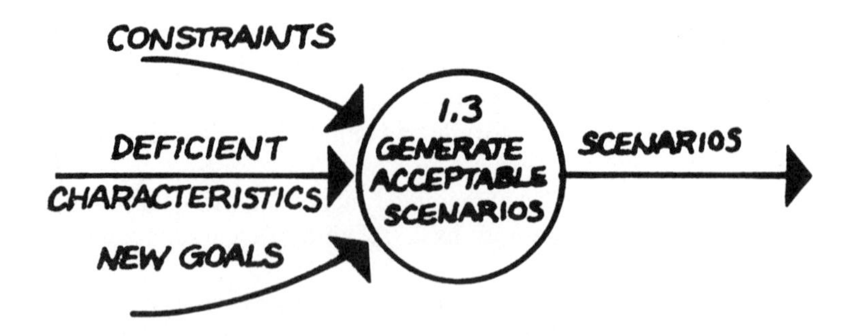

Figure 4.6 Generating acceptable scenarios.

already defined; the other is *constraints.* The output of subactivity 1.3 is the scenario or scenarios. The input (constraints) and the output (scenarios) are defined as follows:

- *Constraints:* time limitations (schedule) + manpower limitations + budget limitations + operational limitations
- *Scenarios:* brief summary of a proposed new system + cost-benefit summary

The meaning of constraints is fairly obvious. The analyst needs to know how much money is available for the project, how much time he can spend developing the system, how many new people can be assigned to the project, and whether he can buy new hardware.[3] However, there are two less obvious observations we should make about the constraints with which the analyst works. First, in the best of all worlds, management provides only preliminary constraints at this stage; for lack of time, the managers probably haven't thought carefully about the limitations they want to impose on budget, manpower, and so on. After they have seen the scenarios, with tentative cost-benefit reports, they may be inclined to revise their constraints. We would expect to see these revised constraints as input to the analysis activity that is discussed in greater detail in Chapter 5.[4]

The second special thing to note about the constraints with which the analyst works is that in the most perfect of worlds, management's constraints will not affect the estimating process by which the analyst arrives at his own version of the schedule, manpower needs, and other requirements. If things work well, management's constraints should represent outer limits within which the analyst's estimates will comfortably fit. In the real world, things don't often work this way: Management's constraints may well be the first stage of a protracted series of bargaining sessions that will result in a *negotiated* schedule and budget.

The reason for having several scenarios from which to select is that this allows users to choose *for themselves* either a quick and dirty solution, a Rolls-Royce solution, or an in-between solution. If the analyst proposes only one scenario, then, in a political sense, it is the analyst's system, so if for any reason it eventually fails (or is perceived to be a failure), it will be the analyst's fault. Note, however, that *all* scenarios are carried forward into the analysis activity, which we will discuss in detail in Chapter 5. All the analyst is doing during

[3]Constraints are also known as operational limitations. For example, management may issue an edict that states, "Any equipment that is acquired for this new system must fit in the existing computer room and must be able to use the existing 208-volt power supply."

[4]This is an example of the feedback phenomenon described when we first introduced the data flow diagram model of the system life cycle in Chapter 3.

the survey activity is identifying several scenarios that are worthy of further examination, but the user's selection is not made at this time.

One comment seems in order before we move on to the final activity of the survey. The project could come to a screeching halt at this point because there may not be any acceptable scenarios. I know of several organizations that periodically go through the survey activity for some of their larger, more critical systems, simply to see whether there are some new solutions worth exploring. Most of the time, the project stops at the end of subactivity 1.3 because management is able to confirm that the present way of doing business, with all its problems and deficiencies, is still the most cost-effective. However, with the kind of dramatic hardware advances taking place today, there is always the chance that a major system could take advantage of improved processing power, storage capacity, telecommunication facilities, and so on.

4.2.4 Subactivity 1.4: Prepare Project Charter

The purpose of subactivity 1.4 is to produce the charter that will guide the subsequent activity of systems analysis; this is shown in Figure 4.7. It establishes project scope, project objectives, fiscal and time constraints, and economic and technical feasibility.

The project charter should include a detailed working plan—a customized project life cycle that defines all of the subsequent activities, subactivities, and deliverables for the project. The *inputs* to subactivity 1.4 are those we have already seen: scenarios, constraints, and new goals. The two outputs from this step are the project charter and a tentative cost-benefit report for each of the scenarios identified in subactivity 1.3. These two outputs are defined as follows:

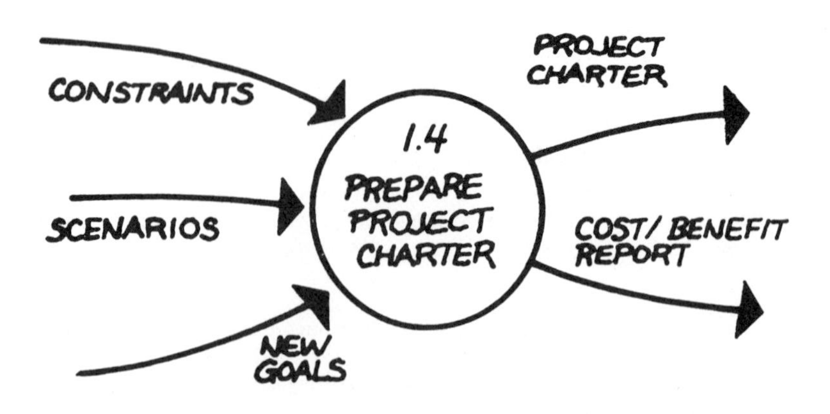

Figure 4.7 Preparing a project charter.

- *Project charter:* project abstract + statement of goals and objectives + customized project life cycle + schedule consraints + technical and procedureal constraints + preliminary project scenarios
- *Tentative cost-benefit report:* costs associated with a scenario + financial benefits associated with scenario + (e.g., savings or increased profits)

As defined here, the project charter has six primary components. The *project abstract* should describe, in a concise, tabular presentation, such particulars as the name of the project; the responsible user; the business area to be studied—specifically, the business boundaries within which the project team must work; the starting date of the project; the target delivery date; the original budget allocation; the project manager responsible; and a short (one- or two-sentence) statement of the purpose (or mission) of the project. During preparation of the project abstract, feedback and iteration should occur between management and the systems analysts. For example, the analysis may propose a target delivery date, and management may negotiate a different date.

A second component of the project charter is the *statement of goals and objectives* which is derived from the new goals that were identified in subactivity 1.2; it is a summary of the functions to be implemented, the deficiencies to be remedied, and the features to be added or modified. In addition, the project charter should include a data flow diagram of the life cycle to be followed for the project, together with the deliverables for each of the activities. In most cases this will be a *customized project life cycle,* tailored from the life cycle presented in Chapter 3.

The fourth component of the project charter is *schedule constraints.* Based on the constraints imposed by management, the project charter should contain an annotated version of the top-level, customized life cycle data flow diagram (similar to Figure 3.4) with key dates and delivery constraints indicated in a PERT format. Note that in order to prepare this customized, annotated DFD, you must have decided whether to use a radical or conservative top-down approach: If a conservative approach is taken, the dates marked on the life cycle will represent deadlines for *completely finishing* the activities of systems analysis, design, and so forth, presumably indicating that the activities will be conducted in serial fashion. In addition to schedule constraints, there are *technical and procedural constraints,* which are also derived from the constraints imposed by management; that is, the project charter should contain a list of the key throughput and volume parameters together with any known restrictions applying to hardware configuration, security provisions, existing system interfaces, environmental conditions, and implementation conventions.

The sixth component of our data dictionary definition of project charter is *preliminary project scenarios.* These are simply the scenarios that were identified in subactivity 1.3. As we noted earlier, the scenarios are documented with a high-level sketch, preferably in DFD format. The purpose is simply to give an overview of possible implementations that could, if carried through to completion, meet the objectives within the stated constraints.

4.3 TOP-DOWN CHOICES FOR THE SURVEY ACTIVITY

As we saw in Chapter 3, we can approach a systems development project in an ultraradical fashion, a moderately radical fashion, a moderately conservative fashion, or an ultraconservative fashion. Indeed, each of the activities, including the survey activity, can be approached in a conservative or radical fashion. In most cases I would argue against selecting the ultraradical or the ultraconservative approach, but both extremes deserve to be discussed.

With an ultraradical approach, the systems analyst would simply skip subactivities 1.1, 1.2, and 1.3 altogether and would simply guess at the items in the project charter. He would plan on a formal performance of all these steps in the systems analysis activity. Obviously, this might be acceptable for very small, informal projects such as those in which the user dashes in and says, "Quick! Write me a program that can go through our master file and produce a report of all left-handed customers! I gotta have it by tomorrow morning!" For anything other than such small projects, however, the ultraradical approach is not recommended.

The moderately radical approach to the survey involves a *superficial* effort to perform subactivities 1.1, 1.2, 1.3, and 1.4 but a *formal* commitment to delve more deeply as the project proceeds. One can imagine, for example, a one-year project that begins with a one-week survey: One week may be sufficient to demonstrate that there is *some* justification for developing a new system, but it is obviously a very superficial justification. The significant point of the moderately radical approach is that it provides the formal recognition of and commitment to the need to document deficient characteristics, new goals, and scenarios in the systems analysis activity.

It should be clear that the survey activity is, in a sense, a minianlaysis and that although the subactivities in Chapter 5 may *look* different from the subactivities of the survey, they accomplish much the same thing. From a political viewpoint, however, the survey activity may have to be accomplished quickly and inexpensively to help justify the larger investment of time and money necessary during the systems analysis activity.

A more conventional approach to the survey activity would be the moderately conservative approach, which would involve a complete performance of subactivities 1.1, 1.2, 1.3, and 1.4, but with no intention of repeating them later in the project. Indeed, most projects seem to opt for the moderately conservative approach, either consciously or unconsciously, perhaps because it would be embarrassing to reexamine the question of cost-effectiveness halfway through the project!

Finally, there is the ultraconservative approach, which involves a thorough study of the user environment. Instead of tentative cost-benefit figures, the ultraconservative approach would produce final figures; instead of tentative scenarios, the ultraconservative approach would produce final scenarios. This means that the ultraconservative survey produces all of the products that are normally associated with systems analysis; it merges all of the subactivities described in this chapter with all those described in Chapter 5. It's hard to imagine why anyone would want to do that unless the feasibility of the project was a foregone conclusion and management wanted the systems analysts to get on with the detailed work.

Regardless of the approach adopted, note that the amount of time spent on the survey does not necessarily have anything to do with its degree of thoroughness. Particularly in a large organization, months can drag by while the systems analyst goes from department to department to conduct interviews with appropriate high-level users.

4.4 PROBLEMS TO ANTICIPATE

As the foregoing discussion hinted, projects don't always go according to plan. If you're the systems analyst or project manager responsible for the success of the project, you should anticipate potential problems during the survey activity. First, it is possible that decision makers in the organization will be in so great a hurry to get on with the systems development effort that the survey may be skipped entirely. This is most likely to be the case if the user announces in a loud voice, "I want a new payroll system, using DB2, and I want it by January first!" If this happens to you, make sure that you get the user's demand in writing, so that nobody will blame you for starting a project that is later shown to be a waste of money.

Second, it is possible that the user may be unable or unwilling to commit to tangible benefits. In the worst case, you'll have a user who says, "I'm sure we would save money if we computerized, but don't ask me how much." Or your user might say, "The new system will give me better information with which to manage my department." If this occurs, keep three things in mind:

- If you and all of the affected users ultimately report to a very practical business person, he may well interpret user fuzziness to mean that there is no justification for the new system and may believe that if you can't document the benefits of the proposed new system, it doesn't deserve to be built.

- If the new system is being requested by a user who is an entrepreneur, you shouldn't push too hard for tangible evidence of costs and benefits. However, be certain that this type of user understands that he's paying the bill for the system. The point is that the entrepreneur makes a dozen "gut" decisions a day, few of which can be justified with hard figures. His marketing strategy, his pricing strategy, his decision to build a new plant or create a new product—all are actions determined by some kind of sixth sense that either works (if he continues to remain in business) or doesn't (in which case he won't be around to ask you to build a new system). To him, your computer system is just another item about which he has a strong hunch.
- Even if the user can't demonstrate tangible costs and benefits, so what? It simply means that the project is a luxury rather than a necessity, and as long as everyone understands that, things will be okay. Politically, it may be wiser to avoid publicizing the lack of justification for the project if your user is a high-level VIP in the organization.

A third problem to be alert to pertains to time: Too much time may be spent on the survey. We commented on this earlier: If an ultraconservative approach is taken, the survey often turns into a full-scale systems analysis. However, the project leader often doesn't consciously decide on an ultraconservative approach; it just gradually happens. One reason for this is that many projects begin without addressing seemingly simple questions: How do we know when we've finished the survey? How do we know when we've done enough studying of the present environment and enough brainstorming about possible scenarios to produce a respectable cost-benefit report? The practical real-world answer is, you finish the survey when you run out of time. However, the deadline should be based on the assumption that the survey will consume approximately 5 percent of the total time of the project. Mangement should decide as early as possible—virtually on the first day of the project—when it wants the survey to be completed; that decision will determine the level of detail to which the systems analyst can investigate.

4.5 SUMMARY

In the mid-1960s, I devoted two years of my life to a project whose charter was either nonexistent or hidden from view. The first-level project leaders (of whom I was one) and the programmers and systems analysts thought that the project's charter was to use (and invent) state-of-the-art computer hardware and software technology for introducing computers into hospitals; top management understood that the project's charter was to develop a working computer system, install it in a hospital, and begin generating revenues within two years. Precisely two years after the project began, the conflicting viewpoints became apparent. The project was canceled—abruptly.[5] For several years afterward, I chalked up this unpleasant experience to youth and naiveté, but gradually I began to realize that this experience was not only a common one but one that will continue throughout the rest of this century. In early 1982, a colleague described a project (and its messy demise) that differed from my own experience only in that (a) the application was banking instead of hospitals, (b) the location was Texas instead of Massachusetts, and (c) the organization was a different *Fortune* 500 company. Like mine, the project had no charter.

The discussion in this chapter has assumed that you won't make this fundamental error and that your primary concern will be developing a proper feasibility study to allow the project to continue. The notion of *iteration,* introduced in Chapter 3, is crucial here: A return-on-investment calculation done at this early stage should be reviewed again and again throughout the project; later review of the charter and overall feasibility could cause it to be canceled despite absolutely perfect progress in the activities of systems analysis, systems design, and programming.

Of course, in the most common data processing projects, the project charter is developed only once, and the feasibility study is conducted only once. In such projects, the critical activity then becomes systems analysis—the job of determining precisely what the user wants the system to accomplish, regardless of how it eventually gets implemented. This is the subject of Chapter 5.

[5]Details of this project are presented in my book, *Design of On-Line Computer Systems* (Englewood Cliffs, N.J.: Prentice-Hall, 1972), Chapter E.

5

Activity 2: Analysis

5.1 INTRODUCTION

The second, and in many ways most important, activity in a software development project is *analysis*. If systems analysis is done well, a mediocre design can be tolerated, and the code can be written by trained chimpanzees. But if systems analysis is done poorly, the best design and the best code will simply allow the project to arrive at a disaster sooner than would otherwise have been possible. Because it is crucial that analysis be done properly, I will devote a considerable amount of attention to it (indeed, significantly more than will be devoted to the other activities).

The primary purpose of the analysis activity is to transform its two major inputs, the project charter and user policy, into a ''structured specification'' that is used as the input for the design activity, for generation of acceptance test data, and for preparation of the user's manual. Figure 5.1 shows the context of analysis as a subset of the overall life cycle that was illustrated in Figure 3.4. Note that Figure 5.1 shows that the primary output of the analysis activity is the structured specification. As we will discuss in this chapter, the structured specification consists of several components:

- A set of leveled data flow diagrams
- A data dictionary
- One or more entity relationship diagrams

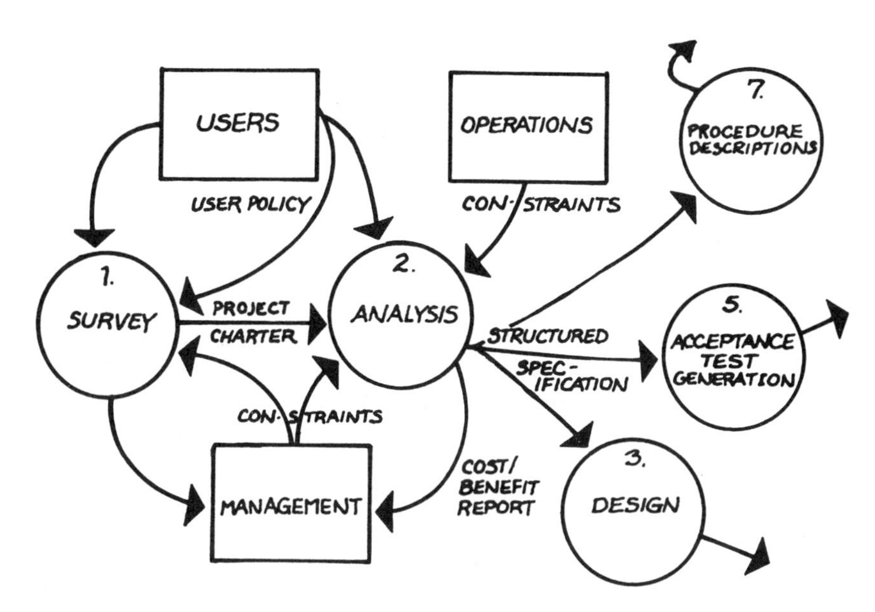

Figure 5.1 Context of analysis.

- Process specifications for all of the bottom-level functions
- A description of the physical constraints that users, managers, and operations personnel wish to impose on the implementation of the system

It is important to recognize that the structured specification represents a *model* of the business; as such, it may prove to be as valuable a product of the project as the eventual computerized system. On several projects that have used structured analysis, users (and managers) at all levels in the organization have been heard to exclaim, "So *that's* what our business is all about!" Indeed, even in the extreme case in which the project is scrapped at the end of the systems analysis phase, a useful product will have resulted.[1]

If the structured specification is a valuable product in and of itself, should it be kept up to date after the project is finished? So far in our discussion of the project life cycle, there has been an unstated assumption that all of the products of each activity are simply intermediate steps in the development of the ultimate product, the installed system. But at least in this case there is the possibility that the structured specification will not only be passed on to the

[1]This implies that the potential usefulness of the analysis model should be included as one of the "benefits" of the project in the tentative cost-benefit report produced by the survey activity. It also implies that structured analysis can be used, alone, as a means of carrying out strategic planning, or business modeling, even when nobody has any intention of automating any part of the business.

design activity and the acceptance test generation activity but will also be sent directly back to the user.

Of course, a model of the user's business won't be of much use if it's not kept up to date—and since the user's business will presumably change over time, some effort will have to be invested in maintaining the model. The effort should be modest, because the structured specification, as we saw in Chapter 2, is by nature easy to modify. But who should be responsible for maintaining the model: the user himself? the project team? Neither of these choices is likely in most organizations, as the project team dissolves at the end of the development effort and the user is too busy (or uninterested) to keep the model up to date. Hence the task may fall on the shoulders of the maintenance programmers—the same individuals who are responsible for modifying the code produced by the project team. In fact, it makes good sense for the maintenance programmers to accept this responsibility. The easiest way of determining where and how to modify the code of a production system is to *begin* with the analysis model: The maintenance programmers and the user can discuss the nature and the impact of the change with the structured specification as the basis for their discussion. The specification can then be changed appropriately, and the change can then be reflected in the code itself.[2]

All of this emphasizes the crucial importance of the structured specification. It has intrinsic value to the user and should be maintained for that reason alone; it forms the basis for rational maintenance of the production system and thus should be maintained along with the code. And as we will discuss at length in the remainder of this chapter, it forms the basis for the actual *design* of the system and becomes the baseline against which the adequacy of that design is measured.

Before we look into the details, we must understand how much of the user's business area should be studied. As we discussed in Chapter 4, the domain of study should be determined in the survey activity and should be documented as part of the project charter. During the survey, our primary concern is that the scope of the project be defined broadly enough to encompass all the areas of the user's business that might be affected by a new system. In the analysis activity, however, our concern is often just the opposite: We need to make

[2]Although most people would agree with this point of view, it is nevertheless rare for the structured specification to be kept up to date. The reason for this is that the graphical component of the model—the data flow diagrams, entity relationshp diagrams, and so on—are usually drawn manually, and nobody wants to take the time and effort to redraw the diagrams if the user's policy changes. The solution to this problem is an *analyst workstation* that provides automated support for *all* the components of the model, including the data dictionary and the process specifications. Approximately two dozen vendors are currently offering such workstations, usually based on the IBM PC; however, only about 1 percent of the systems analysts in the United States had such technology available to them in 1987. It is expected that by 1990 at least 10 percent of the professional systems analysts will be using such tools, and five or six years later it will be a standard tool for at least half the population of systems analysts.

sure that the systems analysts do not study areas of the user policy that the project charter specifically marked as "off limits."

5.2 THE DETAILS OF ANALYSIS

Historically, the systems analysis phase of a project often received very little management guidance or intervention, aside from an intermediate determination of when it should be finished. As we will see in this chapter, there is in fact a great deal to manage in the analysis activity. There are numerous subactivities, each with inputs and outputs, and they can all take place in parallel, which certainly requires a great deal of supervision and control. The details of analysis are shown in Figure 5.2.[3]

Before we discuss the subactivities of analysis, take a closer look at Figure 5.2, the overview diagram. There are four general points to note. First, note that several of the activities in Figure 5.2 may take place simultaneously. There is always a temptation to read the context graphic as if it were a flowchart, but there is no reason why *all* of subactivity 2.1 has to be finished before subactivity 2.2 begins. Indeed, as we will see later in this chapter, there are often very good reasons for overlapping the activities. Second, realize that *all* the analysis activities may be overlapped with design, which, in turn, may be somewhat overlapped with implementation. Once we begin modeling the user's new system, it may be appropriate to commence some design activities immediately—for example, hardware orders may need to be placed one to two years before installation of the system.

Third, it should be emphasized that the conduct of analysis, as discussed in this book, involves an enormous amount of low-level *detailed* work, work that would rarely be assigned to, or accepted by, senior-level software systems analysts, who are normally the *only* people assigned to the project during the analysis phase. In conventional projects, low-level details are almost always postponed until the coding phase, when junior programmers are hastily brought in to reinforce the senior-level analysts. In structured systems development, however, many of these low-level details are considered details of *analysis* and as such should be dealt with during the analysis activity proper.[4] The fourth

[3]Readers familiar with the first edition of this book or with such classic structured analysis books as DeMarco's *Structured Analysis and System Specification* (1978) may have noticed that Figure 5.2 does not involve any modeling of the user's *current* system. This is discussed further in Section 5.3.

[4]Managers should be aware that if a project follows the methodology outlined in this book, junior-level staff may be required at a much earlier stage than in conventional projects. Indeed, perhaps senior-level people should do senior-level work *throughout* all activities of analysis, design, and coding. Similarly, junior people could be assigned low-level tasks throughout the project; for example, writing low-level process specifications for the trivial functional primitives, drawing structure charts for small parts of the system, and performing low-level coding assignments.

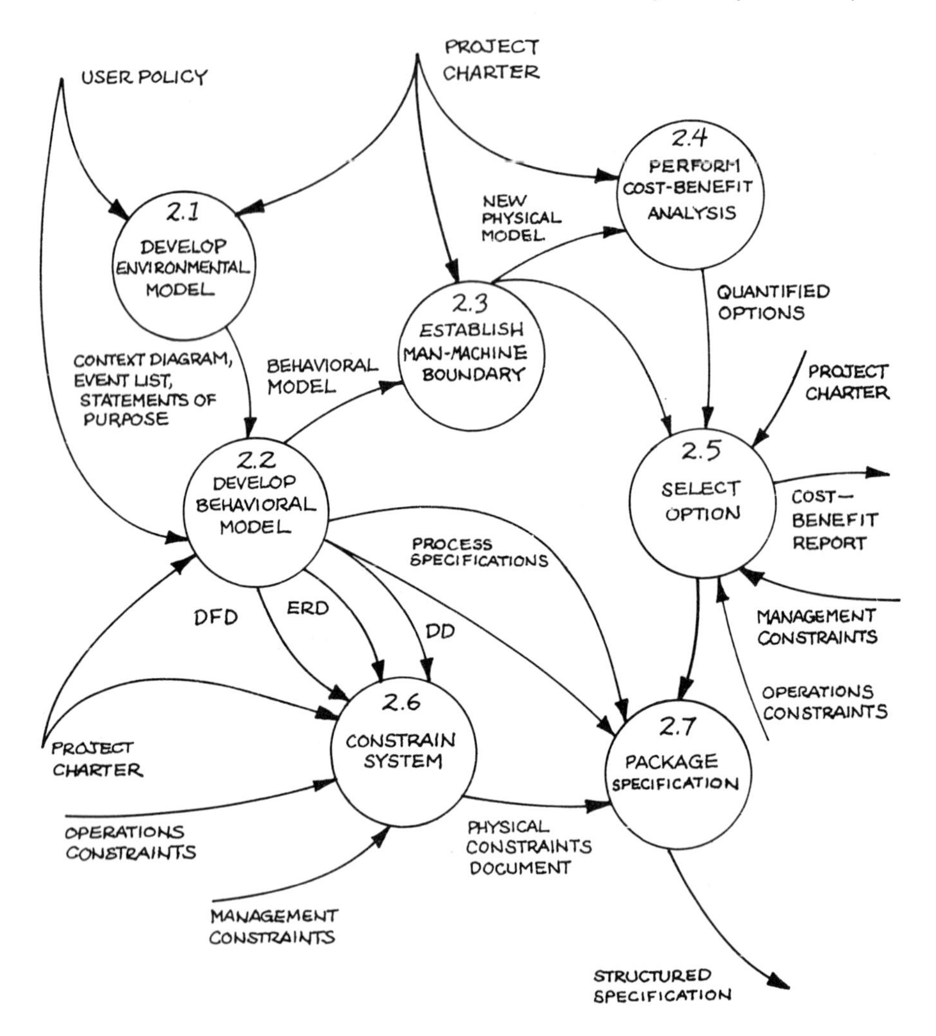

Figure 5.2 Overview of analysis.

point to be noted is that walkthroughs and formal reviews are appropriate at the end of each of the activities shown in Figure 5.2. In addition, walkthroughs and reviews can take place at each *level* of decomposition in the modeling processes of subactivities 2.1, 2.2, and 2.3. That is, one does not have to wait until the entire behavioral model is complete in order to have a useful walkthrough; the top two levels of the model will probably be enough to justify one.

5.2.1 Subactivity 2.1: Develop Environmental Model

The first two subactivities of analysis involve the development of an *essential model* of the user's new system.[5] An "essential" model, as the term implies, is a model of the *essence* of the user's system—a model of the things that must be done regardless of the technology that is eventually selected to implement the system.

The essential model consists of two major pieces: the environmental model (which is produced in subactivity 2.1) and the behavioral model (which is produced in subactivity 2.2). The environmental model is, as its name suggests, a model showing how the system interacts with the outside environment; the behavioral model is a model indicating what the system must do to interact satisfactorily with that environment.

Development of the environmental model is shown in Figure 5.3. It has two inputs: the project charter and user policy. And it produces two outputs:

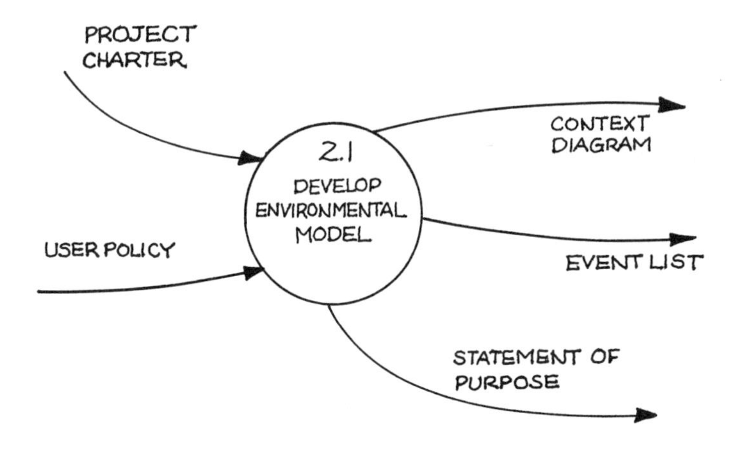

Figure 5.3 Developing the environmental model.

[5]In many of the classical textbooks on structured analysis, such as those by DeMarco (1978) and Gane and Sarson (1979), the term *logical model* is used in the same way as *essential model* in this text. Many practitioners of structured analysis have found, over the past several years, that the term *logical* created a great deal of confusion, especially in discussions with users. The question often asked by a user was, "If a model isn't logical, does that mean it's illogical?" More recent books, such as the one by McMenamin and Palmer (1984), now use the terms *essential model* and *implementation model* instead of *logical model* and *physical model*, respectively.

a *context diagram* and an *event list.*[6] A context diagram is a special case of the data flow diagram discussed in Chapter 2: It consists of a single bubble, or process, representing the entire system. The data flows shown in the diagram represent flows of data between the system and the "terminators"—the people, groups, departments, organizations, and other systems with which this system communicates. A typical context diagram is shown in Figure 5.4.

The event list is a simple textual listing of the "events," or stimuli in the external environment, to which the system must respond and an indication of the person or system that initiates the event. In many cases the system becomes aware of the occurrence of an event because of the arrival of some

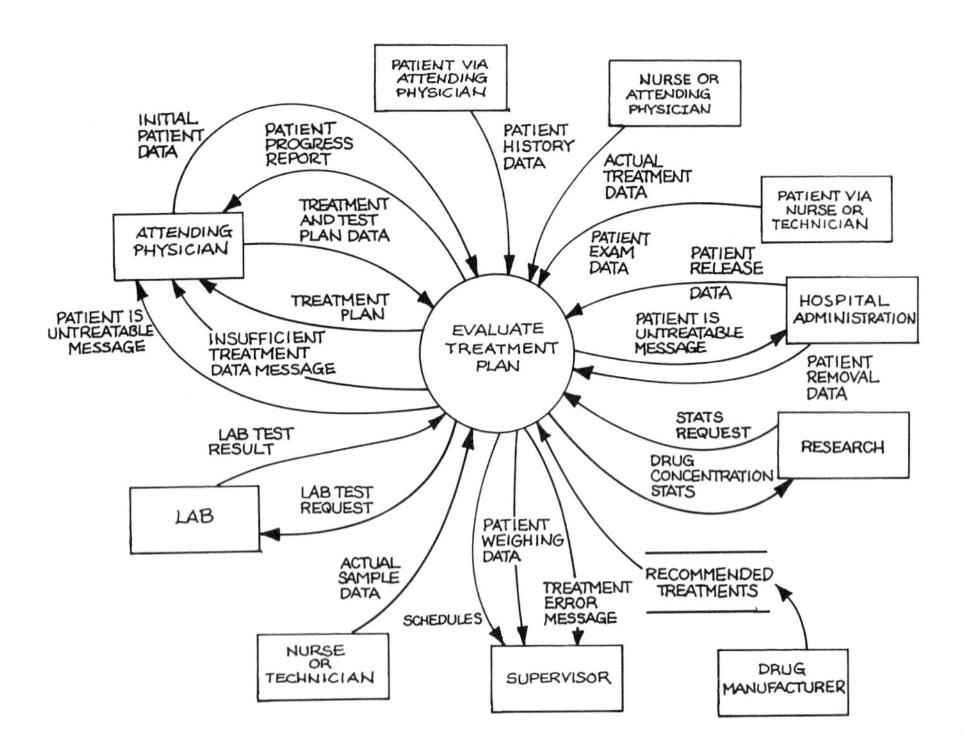

Figure 5.4 A context diagram.

[6]Some project managers insist on a third component in the environmental model, a concise statement of the project's purpose. This is a restatement and amplification of the information already provided in the survey activity's project charter. However, since many of the systems working on the project may not have been involved in the survey activity and may not have read the project charter, it is often politically useful to go through this again to ensure that everyone is on the same "track."

data shown as a data flow on the context diagram[7]—for example, a customer order is received, signaling the occurrence of the event "customer places order." Some events, however, are "temporal" in nature; for instance, an event might be "A weekly report of all transactions is required." Here is a typical event list, associated with the context diagram shown in Figure 5.4.

EVENT LIST FOR THE HOSPITAL SYSTEM

1. Patient joins Nephrology Service.
2. Doctor requests scheduling of initial treatments.
3. Patient signs release form (or not).
4. Supervisor needs test schedule.
5. Lab produces test results.
6. Patients need to be weighed.
7. Treatment schedule required by supervisor.
8. Patient treated.
9. Scheduled treatment missed.
10. Researcher requests drug concentration statistics.
11. Doctor schedules treatments and tests for patient.
12. Sample taken from patient.
13. Patient leaves Nephrology Service.
14. Patient weighed.

Note that the environmental model does not tell us anything about the behavior of the system. Its purpose is simply to document the boundary between the system and the environment and to illuminate the data that must be processed, the data that must be produced, and the outside entities that either provide data to the system or consume data from the system.

5.2.2 Subactivity 2.2: Develop Behavioral Model

Subactivity 2.2 is the heart of systems analysis; this is where the major work of developing data flow diagrams (DFDs), entity relationship diagrams

[7]However, the converse is not true: a data flow coming into the system is not necessarily the indication that an event has occurred. For example, an event in the environment might create some input to the system along data flow A (which is how the system becomes aware of the event). In order to respond to the event, the system may request some input (perhaps from a different terminator) along data flow B. Thus any data flow entering the system will either be an indication to the system that an event has occurred or will be required by the system to produce the response(s) to an event.

(ERDs), process specifications, and the data dictionary (DD) are developed. All of these comprise the *behavioral model,* the model of what behavior the system must exhibit in order to deal with its environment satisfactorily. The behavioral modeling activity is shown in Figure 5.5.

For a simple system, it may be possible to create a single DFD and ERD (and associated DD and process specifications) after interviewing the users. However, most systems are sufficiently complicated that we will end up with several *levels* of DFDs (typically three to five levels) and a total of hundreds of bottom-level bubbles (one of my consulting clients is working on a system with 10,000 bottom-level bubbles!). Thus the systems analyst faces several questions: Where do I start? How do I decide what the bubbles should be in the top-level DFD? How do I decompose each of those bubbles into lower-level DFDs?

Most systems analysts assume that the best way to proceed is in a strictly top-down fashion. They begin with the single bubble shown in the context diagram and "explode" it (decompose it, break it down) into a top-level DFD, each of the bubbles in the top-level DFD (which is usually known as "Figure 0")

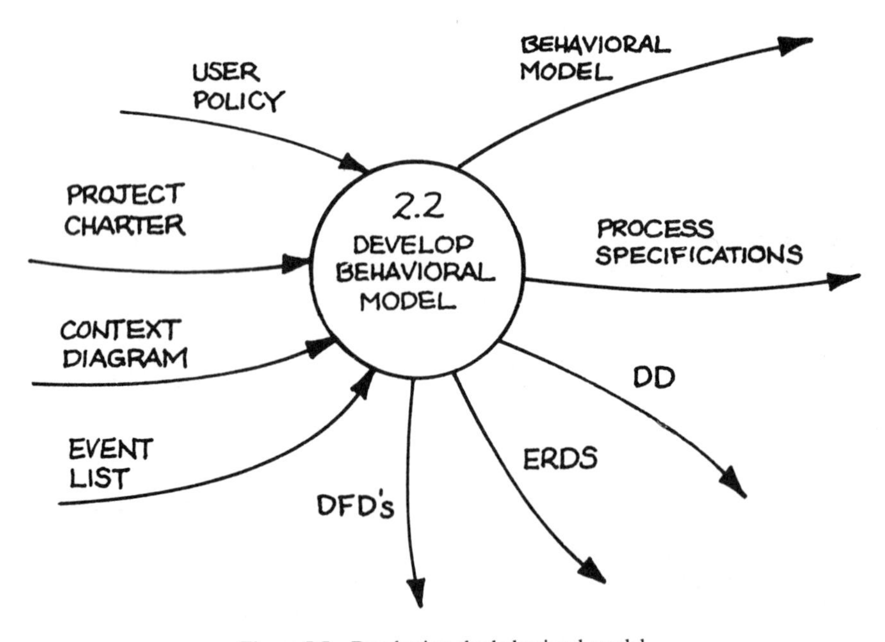

Figure 5.5 Developing the behavioral model.

is broken down further, and so forth. However, this procedure suffers from three common dangers:

- *Paralysis:* In some cases the systems analyst stares at a blank sheet of paper and is simply unable to imagine what the top-level processes in Figure 0 should be.
- *The "six-analyst" phenomenon:* If there are six systems analysts working on the project, there is an unfortunate tendency to draw Figure 0 with six bubbles in it, so that each analyst can continue to work on his own. Unfortunately, this partitioning of the system into six pieces may have little or nothing to do with the "natural" organization of processes in the system.
- *Physical bias:* If a new system is being developed to replace an existing system, there is a tendency to organize the top-level processes in Figure 0 in exactly the same way that the current system is partitioned or according to the organization's current structure. If the current system is entirely manual, for example, the top-level "processes" may be represented by groups of people (e.g., accounting department, engineering department, administrative services department, etc.). This may or may not represent good partitioning in the new system.

To avoid these problems, we suggest that the systems analyst use the event list to produce a "first-cut" set of DFDs, which can then be reorganized into appropriate leveled DFDs for presentation to the user. This process, known as *event partitioning,* consists of the following steps:

1. Draw a bubble, or process, for each event on the event list.
2. Name the bubble by indicating the response(s) that the system must make in order to respond to the event.
3. Draw in the necessary inputs and outputs that the bubble requires in order to respond properly. Also draw in the data stores that represent communication among the bubbles.

This approach produces a large, clumsy-looking DFD that is almost certain to confuse the users, for if there are 45 different events, the first-cut DFD will have 45 bubbles in it. Thus the systems analyst must go through an *upward leveling* procedure to group closely related bubbles and create higher-level bubbles (which form higher-level DFDs); this is often accompanied by some *downward leveling* (decomposition) activities to break complex bubbles into more easily understood, lower-level DFDs.

5.2.3 Subactivity 2.3: Establish Man-Machine Boundary

After the essential model has been completed, the next subactivity is establishing a man-machine boundary. In subactivity 2.3 our task is to identify a number of *possible* man-machine boundaries, from which one will eventually be selected. this is shown in Figure 5.6.

The inputs to subactivity 2.3, as we can see, are the behavioral model generated by subactivity 2.2 and the project charter generated by subactivity 1.4 in the survey; the output is a *set* of new physical models, each of which is defined as follows:

- *New physical model:* leveled data flow diagrams and data dictionary of the new system, which have been physicalized only to the extent of describing which processes and data elements will be automated (those contained within a computer system) and which will be manual (those performed by human beings or within their purview).

Note that the alternatives considered here usually do not involve definitive hardware selection; it should not be necessary for the systems analyst to say, "One of the options considered is a DEC VAX 11/780 computer." Instead, we simply identify various *classes* of hardware. For example, the systems analyst might state, "One option is to put the entire system into a single mainframe computer; another option is to encapsulate each bubble in the DFDs in its own microcomputer." And the systems analyst should be concerned about the *extent* of automation—precisely which portion of the overall system will be carried out by humans and which by computers.

Note that the project charter is one of the inputs to this subactivity. Why? Simply because subactivity 2.3 is, in a sense, a more detailed version of sub-

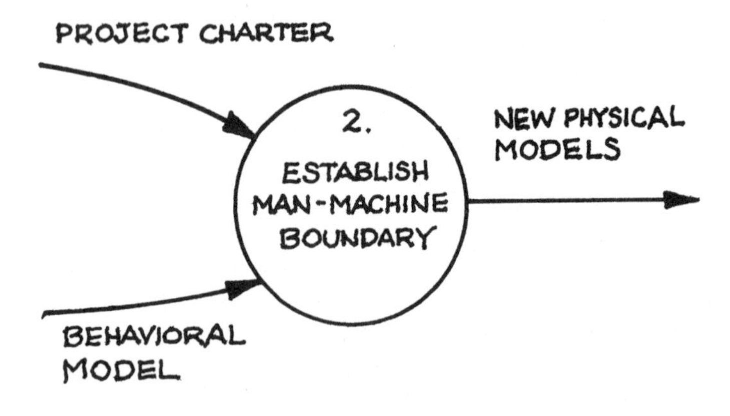

Figure 5.6 Establishing the man-machine boundary.

activity 1.3, in which scenarios for a possible new system were first proposed. Thus the project charter provides the analyst with some earlier views of possible implementations of the system; either he can confirm their validity, based on the new logical models, or he can reject them and propose entirely different new physical models.

5.2.4 Subactivity 2.4: Perform Cost-Benefit Analysis

Having generated several possible new physical models in subactivity 2.3, the next step is to quantify them—that is, to attach costs and benefits to each option. This is shown in Figure 5.7.

As we can see, the inputs to subactivity 2.4 are the new physical models and the project charter produced in subactivity 1.4 of the survey. The output of subactivity 2.4 is a set of quantified options, defined as follows:

- *Quantified option:* identification of a particular new physical model (such as "the distributed minicomputer option" or "the on-line option") combined with information about costs and benefits in sufficient detail for management to be able to select one option over another

The quantification of the new physical models will usually be in terms of *development costs* (how much will it cost to acquire the hardware and develop the software?), *operational costs* (how much will it cost to run the hardware on a day-to-day basis?), *schedule* (can we get it running by the beginning of the new fiscal year?), and *risk* (if we're buying all the hardware from the El-Shakee Computer Company, what are the chances that the

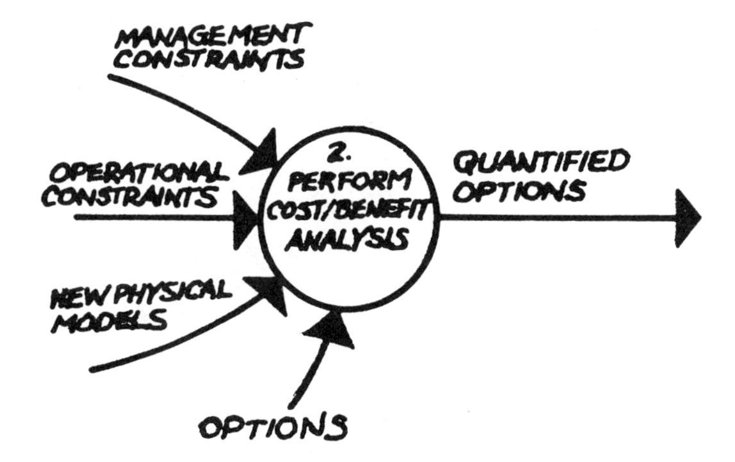

Figure 5.7 Performing cost/benefit analysis.

company will be bankrupt before the equipment is delivered?). Of course, all are based on estimates. But the estimates should be considerably more accurate than they were in the survey. The reason that the project charter is an input to subactivity 2.4 is that *tentative* cost-benefit information was first computed in subactivity 1.3; thus the analyst can take advantage of those preliminary calculations and either refine them or reject them and begin anew.

5.2.5 Subactivity 2.5: Select Option

Once the various man-machine options have been quantified, there remains the separate activity of actually *selecting* an option; that is obviously the function of subactivity 2.5, shown in Figure 5.8.

As we can see, there are four inputs to this activity: project charter, quantified options, operational and management constraints, and new physical models. The two outputs are a cost-benefit report and a selected physical model, defined as follows:

- *Cost-benefit report:* costs associated with the selected physical model plus savings associated with the selected physical model
- *Selected physical model:* leveled data flow diagrams and data dictionary that have been physicalized, but only to the extent of distinguishing between the parts of the system that will be automated and those that will be performed by humans

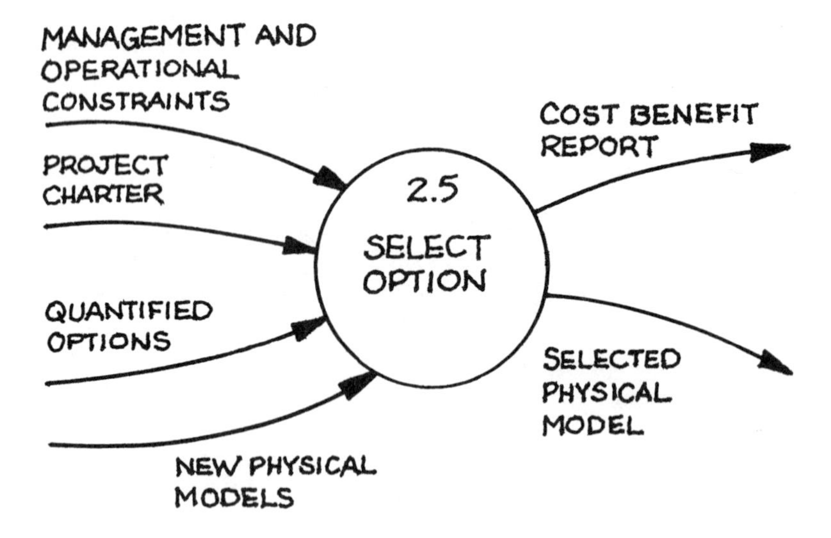

Figure 5.8 Select option.

It may appear somewhat artificial to separate the *quantification* of the various man-machine options from the *selection* of an option, but the separation is valid for at least two reasons. First, in many cases, the choice will be made by management, rather than by the systems analyst who compiles all of the relevant data. Thus the management constraints might well take the real-world form of *control* rather than *data*. Second, the final choice of an option may involve a number of considerations that transcend a simple computation of costs and benefits—for example, more complex cash flow considerations, tax ramifications, or politics of one form or another.

Please keep in mind that the physical nature of the selected physical model is only *barely* physical; the parts of the system that are within the domain of the computer are still expressed in entirely *essential* terms, that is, in terms of the essential data flow diagrams and data dictionary. The ultimate physicalization will be influenced by subactivity 2.6, as we will see shortly, but it will not be fully accomplished until subactivity 3.5.

5.2.6 Subactivity 2.6: Constrain System

All of the analysis activities that have been described so far have been primarily concerned with the *essential* aspects of the new system—its inputs, processing requirements, and outputs—but not the physical means of implementation. Even though subactivities 2.3, 2.4, and 2.5 deal with new physical models, they address only a small part of the physical aspect of the system.

This emphasis on the essential is as it should be. Neither users nor managers should impose their views on the systems analyst or designer when it comes to choosing a particular physical implementation. They may specify constraints of cost and cash flow and so forth, but they should not prescribe the overlay structure, the nature of the database access method, or even the programming language that will be used to implement the system. (Most of the time, of course, they will not *want* to involve themselves in such mundane issues, but it is becoming more and more common for both users and managers to attend one-day "computer appreciation" seminars at which they learn how to write a small program in BASIC on a desktop microcomputer, which makes them instant experts in the software field—and before you know it, they're telling the designers and analysts that *everything* should be programmed in BASIC.)

However, there are always real-world considerations. There are always some *physical* aspects of the system about which the users or the managers feel strongly enough to include in the structured specification. Or, to put it another way, if the analysis and eventual implementation are to be performed by an outside software company, there will always be some physical aspects of the system that the users or the managers feel strongly enough about to put into a formal contract.

The main purpose of subactivity 2.6 is to produce a concise description of the full set of management-imposed constraints and their signficance to the implementers. Since the rest of the structured specification is (or should be) nonphysical in nature, the physical considerations (apart from the man-machine boundary, which was determined in subactivity 2.5) should be confined entirely to the physical constraints document produced by subactivity 2.6.

As can be seen in Figure 5.9, the inputs to subactivity 2.6 consist of items that we have already discussed: management constraints, project charter, data flow diagrams, and the data dictionary. The output consists of the physical constraints document, which is defined as follows:

- *Physical constraints document:* summary of all physical constraints imposed on the proposed new system

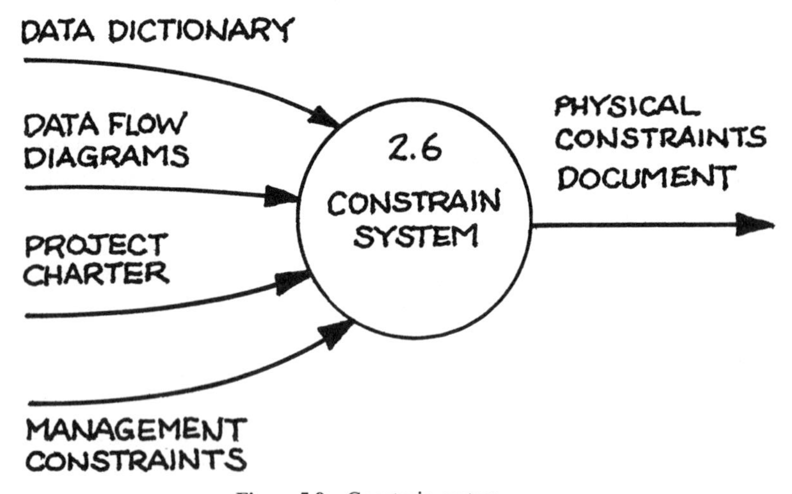

Figure 5.9 Constrain system.

To give you a clearer understanding of the types of items that are likely to appear in a physical constraints document, here are some isolated examples:

- No new hardware may be acquired on the project.
- Hardware must be acquired from vendor A.
- Hardware may be acquired from any vendor *except* vendor A.
- Response time on the system must be less than 3 seconds.
- Mean time between failure (MTBF) must be at least one month.

- Mean time to repair (MTTR) must be less than 10 minutes.
- Report No. 32 must be produced no later than 9:00 A.M. each morning.
- The computer hardware must be capable of operating in a normal office environment, with no special air conditioning or additional electrical wiring.
- 90 percent of all responses to inquiries must occur within 2 seconds.

5.2.7 Subactivity 2.7: Package Specification

The last step in the analysis activity is mostly clerical. Its purpose is to collect and integrate the outputs of the other subactivities into one coherent document. As you can see from Figure 5.10, the inputs to subactivity 2.7 are the physical constraints document, the data dictionary, the process

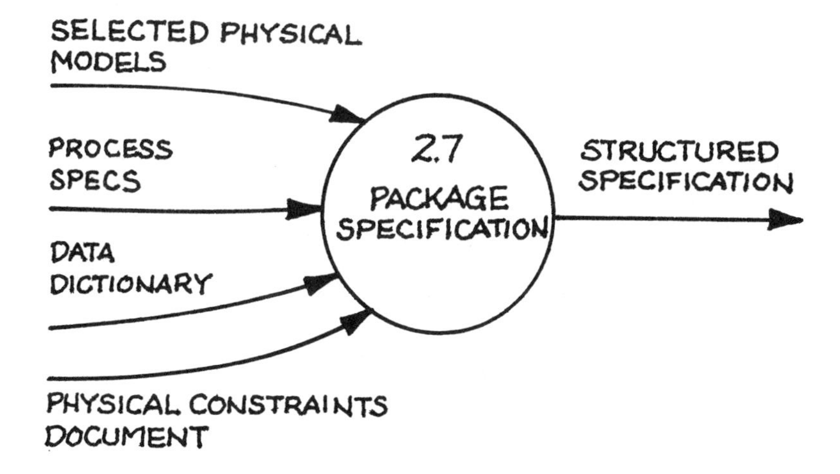

Figure 5.10 Package specification.

specifications, and the selected physical model. The output is the structured specification, which is defined as follows:

- *Structured specification:* leveled data flow diagrams + data dictionary + process specifications + physical consraints document.

A brief narrative, an index, and a boilerplate description of the use and makeup of the structured specification should be added to complete the analysis activity.

5.3 MODELING THE CURRENT SYSTEM

The approach just shown is substantially different from the one proposed in several classical structured analysis textbooks; a more common approach is shown in Figure 5.11. Note that it differs from the activities discussed in Section 5.2 in one important respect: The "classical" approach to structured analysis emphasizes modeling the user's *current* system, while the approach suggested in this book ignores the current system and emphasizes modeling the user's *new* system.

There are three reasons for modeling the user's current system:

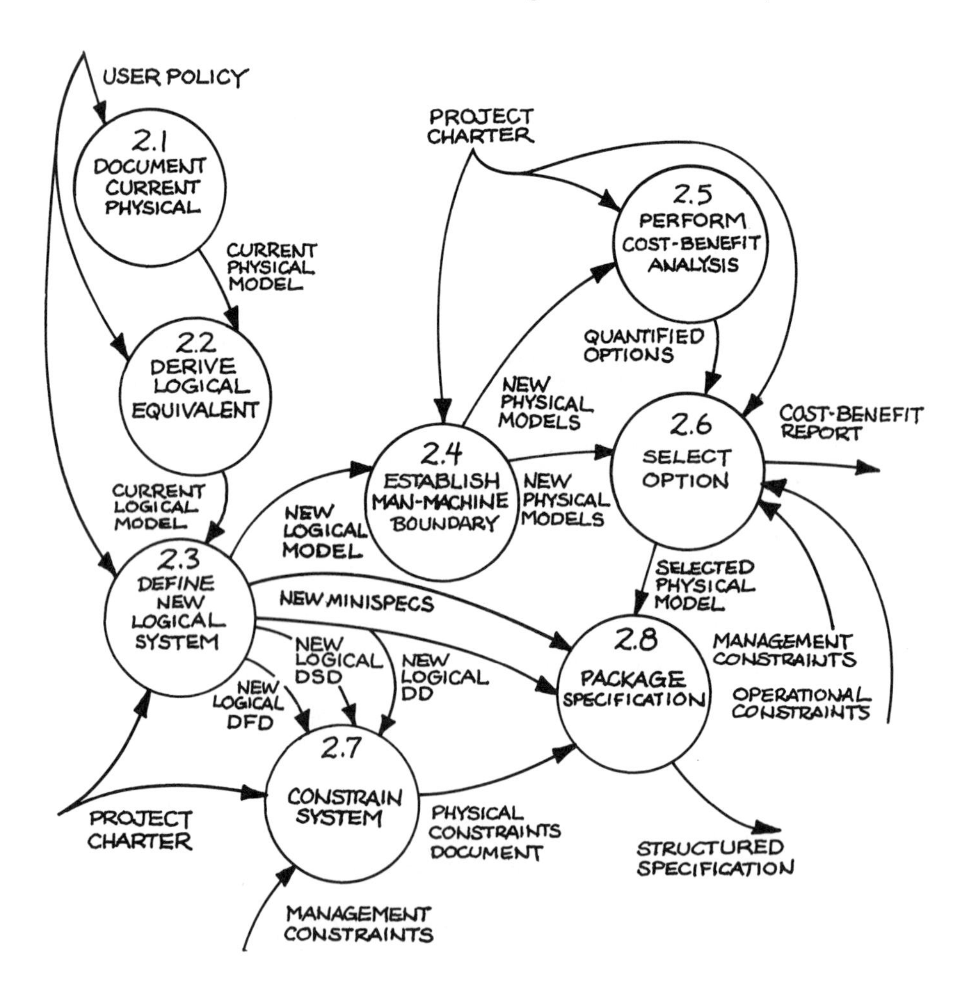

Figure 5.11 A classical view of the analysis activity.

- The systems analyst often does not understand very much about the end user's business. Hence he may be rather nervous about modeling a new system for the user before he understands what the user is doing now. Modeling the current system can be a good learning experience.
- Even if the systems analyst thinks that he is an expert in the user's business, the user may not believe it. Hence it may be necessary for the systems analyst to develop a model of the user's current system just to gain credibility. Also the analyst is ultimately trying to create an essential model of the user's new system, but that model may be so strange, abstract, and unfamiliar that the user will have trouble understanding it. The analyst may have to begin with a model that contains familiar "landmarks" that the user can understand and verify.
- Since most new systems are very similar, functionally, to the systems they replace, there probably won't be much wasted effort documenting the current system's functions. In most cases, 75 to 80 percent of the functions in the user's current system will be implemented again in the new system, along with new functions necessary to satisfy user needs not currently addressed.

Thus the classical approach to systems analysis begins with a detailed model of the user's "current physical" system—usually a collection of manual procedures and automated (computerized) procedures that may or may not be well documented. This is followed by a "cleanup" process that leads to an *essential* model of the user's current system, a model of the "pure" business policy that the user is carrying out, independent of the arbitrary implementation of that policy.

Why, then, have we presented a different approach in this book? Simply because the experience of several hundred projects over the past several years has shown that there are dangers involved in modeling the user's current system. The three primary dangers are these:

- There is a tendency to spend too much time modeling the current system; this has been described by McMenamin and Palmer (1984) as the "current physical tarpit." Since there is such a wealth of information to document about the user's current system, the systems analyst sometimes forgets that the ultimate objective is to develop a new system. After two or three years of modeling the existing system (yes, that's right, *years* can go by while the analyst does this work), the user gets impatient and cancels the project.
- Much of the work modeling the user's current system may turn out to be wasted. Although the *functions* carried out by the user today may

be carried over to the new system, the *implementaion* of those functions is usually quite different. The current system often has a great deal of redundancy; it also contains numerous editing, validation, error-correction, and "transporter" functions that exist only because of the current technological implementation. Many organizations have found that only 25 percent of a current system model is "essential" in nature; 75 percent of the model is concerned with the arbitrary details of implementation and is thus thrown away when the new system is developed. The problem is exacerbated, as my colleague Bob Spurgeon points out, by the fact that the 75 percent that is to be thrown away is usually not blatantly obvious; many implementation details are unbelievably subtle.

- If the systems analyst spends to much time modeling the user's current system, there is a significant danger that he will be "brainwashed" into believing that the *implementation* of the current system is, in fact, the *essence* of the current system. That is, he is likely to specify a new system that replicates many of the arbitrary technological decisions of the old system; the user is even more likely to make the same mistake.

Because of these problems, most veterans of structured analysis now feel that modeling of the user's current system should be deemphasized, and ignored altogether if possible. If such a model must be developed, it should be done only to the extent necessary for the user and the systems to agree that they understand the functions of the current system, so that attention can then be focused on the rquirements of the new system.

5.4 TOP-DOWN CHOICES FOR ANALYSIS

As we saw with the survey, there are a variety of ways to approach the top-down implementation of all of the activities in a software development project. Let us briefly discuss some of the top-down options for analysis.

With an ultraradical approach, one could imagine all of the subactivities discussed in this chapter being based on two or three levels of data flow diagrams and the data dictionary. Typically, *no* process specifications would be developed with such an approach, and the analysis would continue for the remainder of the project. In general, I would advise that you *not* follow the ultraradical approach, since it is difficult (if not impossible) to gain any real understanding of the user's requirements if you examine it to only two or three levels of detail. However, it may be necessary if the user himself is not sure of his requirements, and a prototyping approach is called for.

The moderately radical approach is a bit more tolerable. It presumes that the systems analyst performs all the subactivities described in this chapter to within one or two levels of the bottom of the user's system. One can imagine the systems analyst saying, "Ah, yes, Mr. User, I see that you have a process called VALIDATE CUSTOMER ACCOUNT NUMBER, and I have an idea of what sort of things are involved in that validation." In fact, the validation rules may turn out to be significantly more complex than the systems analyst realizes, but at least he can see that he is one level from the bottom of the leveled DFDs. As we discussed in Chapter 3, there are a variety of reasons why the analyst—or the analyst's boss!—might opt for the moderately radical approach. The key point is that analysis must continue in parallel with the subsequent activities in the project.

A moderately conservative approach would involve the identification of *all* of the bottom-level bubbles in the leveled data flow diagrams. It would involve a complete process specification—in structured English or any other suitable form—of the business policy of each of the critical bubbles, though not necessarily all the trivial ones. And it would involve a data dictionary definition of all data elements down to the field level. For medium-sized and large projects, this would be the preferred approach. It would provide sufficient information with which to produce reasonable cost-benefit reports and to develop an implementation model intelligently. Even though considerably more detail is involved in the moderately conservative approach, it is still evident that the job of analysis has not been completed; some of the details will continue to be developed in parallel with the subsequent activities of the project.

Finally, there is the ultraconservative approach, involving *all* the steps described in this chapter, carried out to the last iota of detail. I recommend this approach with one reservation: By the time the ultraconservative analysis has been finished, some of the information gathered might well be obsolete! And there is always the danger that some of the information gathered is simply *wrong,* due to a misunderstanding between the analyst and the user. So if the ultraconservative approach is chosen, hope that the user is extremely knowledgeable about his requirements and can communicate easily with the analyst and that the user's environment is stable and won't change before final implementation of the system is accomplished.

5.5 PROBLEMS WITH ANALYSIS

What can go wrong with analysis? How could anyone ignore or avoid the simple, sensible steps we have outlined in this chapter? In practice, there seem to be four major dangers. The first is *impatience on the part of management.* In many cases management may not be prepared to spend the amount of time on systems analysis that is indicated by the discussion in this chapter. Indeed, management may not be willing to spend *any* time on systems analysis, since in the opinion of many managers, analysis is simply a period of rest preparatory to getting ready for coding, the "real" work of the project. In fact, the work discussed in this chapter will generally account for between 30 and 35 percent of the time and resources of a typical software development project—and in any form of a conservative project, there will be no visible evidence of progress during this period (other than pages filled with bubbles and boxes). No wonder management gets paranoid! Indeed, there is a great temptation to hide some of the details of analysis in some of the later activities of systems development—and that, of course, is what always used to happen: The junior-level programmers and junior-level users did all of the detailed systems analysis work on an informal basis long after analysis and design supposedly were officially finished!

The second danger is similar to the first and is self-explanatory: *impatience on the part of users.* All of the characteristics described in the preceding paragraph may well occur in conjunction with users.

As analysis progresses, a third danger may arise: *The structured specification may drift away from the project charter.* This is especially common in large projects, in which the survey is conducted by people other than those who conduct the analysis. One of the most common manifestations of this drift is the addition of new functions that the user doesn't even want. Formal reviews involving responsible people outside the develoment team should be held at the end of the analysis activity to avoid this phenomenon of "feature creep."

A last problem to be alert to involves the fact that *technical errors may appear in the structured specification.* There is no doubt that preparing the structured specification can seem a bit overwhelming for a systems analyst when he first attempts it. Thus it is not surprising to see such errors as *infinite sinks* (bubbles with one or more inputs but no outputs), *write-only databases* (databases that are updated but never read), *balancing errors* (child DFDs whose net inputs and outputs don't match those of the parents from which they are derived); and *undefined data elements or databases.*[8]

5.6 SUMMARY

Management of the systems analysis portion of an EDP project is probably more difficult than management of any of the other areas presented in this book. As you can see from the discussion in this chapter, there is a great deal to manage in connection with the numerous subactivities, each with several levels of detail and all of which must be coordinated. And by definition, system analysis is the portion of the project that directly involves the user—so it is

[8]As mentioned earlier in this chapter, a number of automated "analyst workbench" products are beginning to appear in the marketplace. In addition to greatly simplifying the chore of drawing data flow diagrams, the products check for balancing errors, undefined data elements, infinite sinks, and the like. This makes it considerably easier to create error-free specifications.

the most prone to problems of politics, misunderstandings, miscommunication, and so forth.

Though it is certainly true that a good job of systems analysis will greatly simplify the subsequent activities of design and programming, the drawback is that more time and energy need to be devoted to analysis than on projects using the classical approach. And if *all* the tasks outlined in this chapter are completed before any design, coding, or implementation have begun, there is a significant danger of impatience and frustration on the part of the user. Iterative development, discussed in Chapter 3, helps to alleviate these difficulties.

Assuming that the activity of analysis has been done successfully, the next major job is that of developing a blueprint for construction of the automated part of the system—determining how the hardware and software will be organized, how the software will be partitioned into pieces, and how the pieces will communicate with one another. That is the activity of design, discussed in Chapter 6.

6

Activity 3: Design

6.1 INTRODUCTION

As we first discussed in Chapter 2, *design* is concerned with identifying *which* modules and *which* interfaces will best implement a well-specified problem. Figure 6.1 shows the context of design in the structured project life cycle; once again, keep in mind that this is a subset of the entire life cycle that we first saw in Figure 3.4 (repeated for your convenience on the following page).

Figure 6.2 shows an overview of the design activity and is made up of seven bubbles. The first bubble allocates portions of the structured specification (portions of DFDs and data stores) to individual processors; the second bubble allocates portions of the specification to individual tasks within each processor. The third bubble indicates the derivation of a set of structure charts for each task-level portion of the specification. In the fourth bubble, evaluation of the structure chart occurs. The next bubble uses the refined input, data dictionary entries, and process specifications to design the "innards" of individual modules. Database design, indicated in bubble 3.6, uses the data dictionary and the ERDs provided by the structured specification. The final component of the design activity involves packaging the design. These seven components plus inputs and outputs are shown in Figure 6.2.

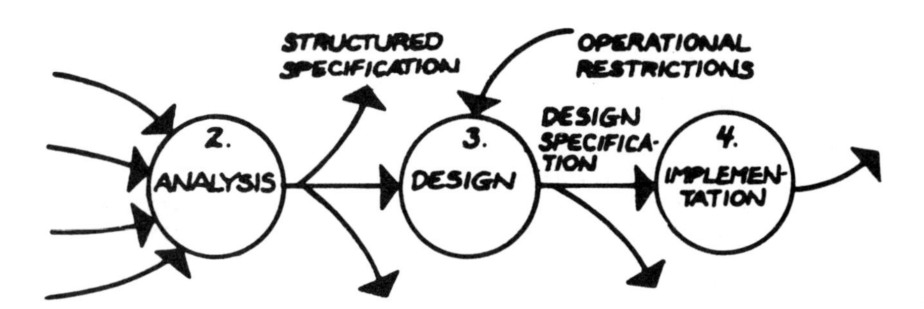

Figure 6.1 The context of design.

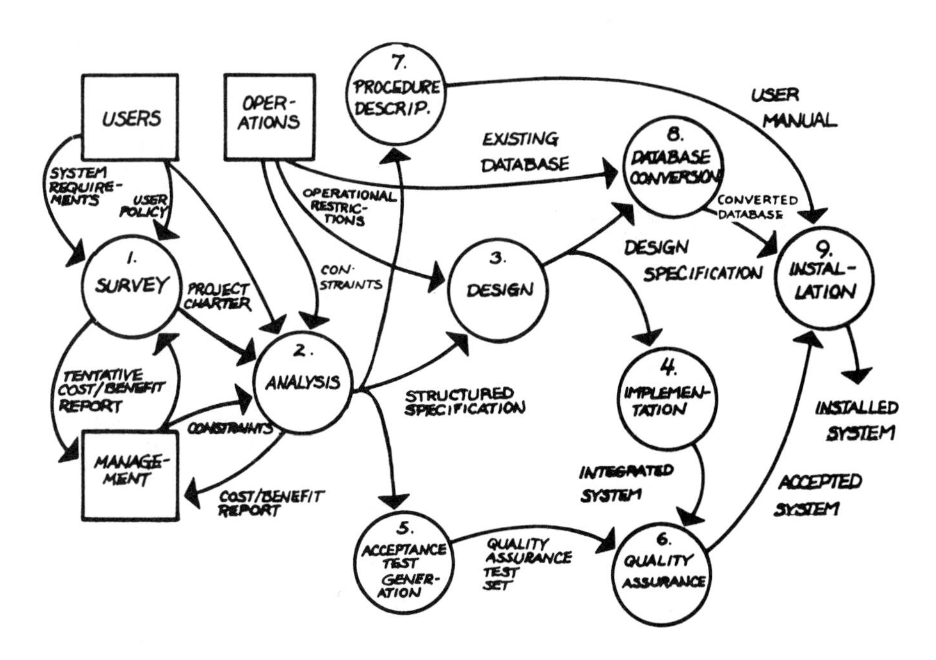

Figure 3.4 Structured project life cycle.

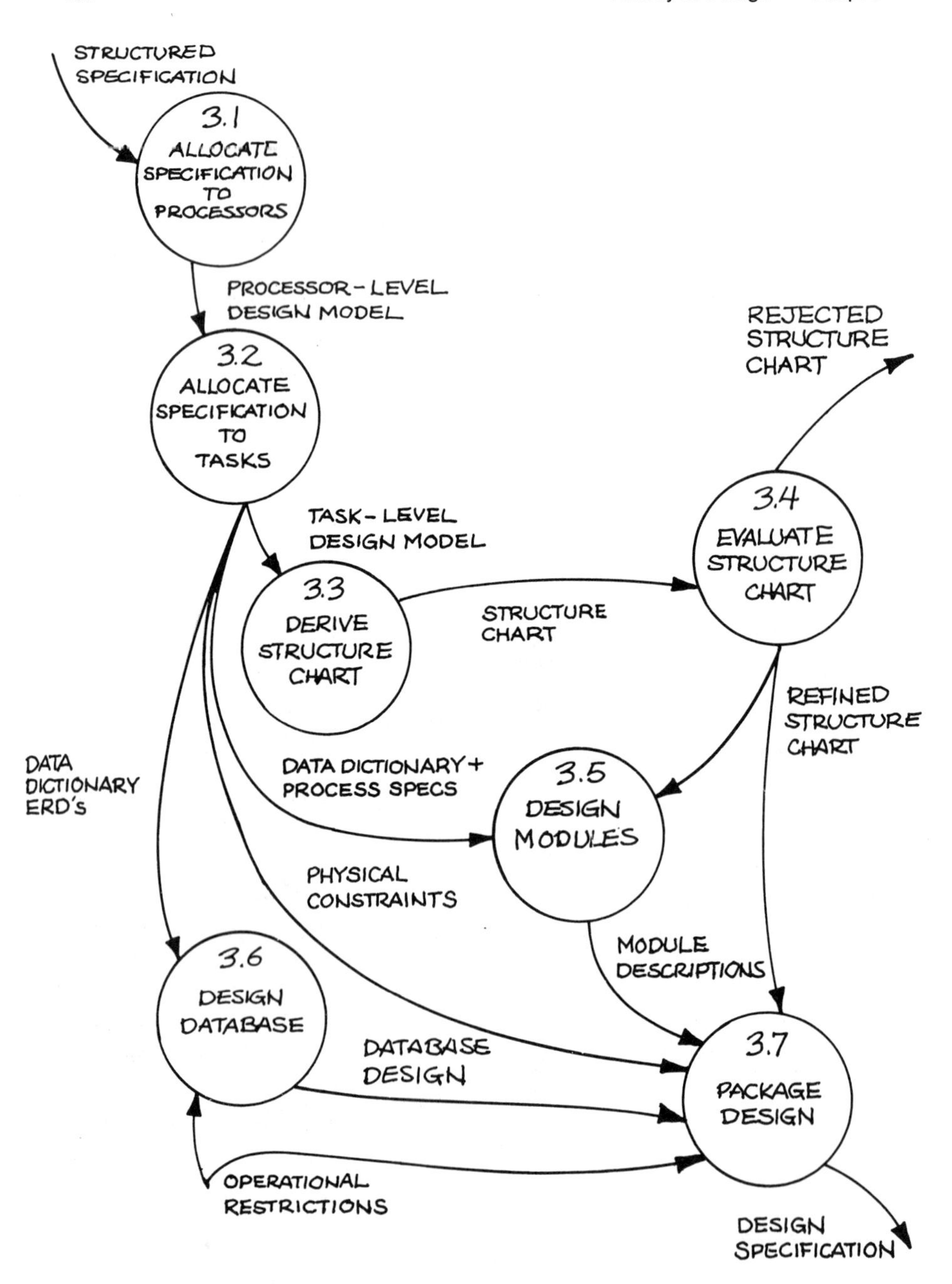
STRUCTURED SPECIFICATION
3.1
ALLOCATE SPECIFICATION TO PROCESSORS
PROCESSOR-LEVEL DESIGN MODEL
3.2
ALLOCATE SPECIFICATION TO TASKS
REJECTED STRUCTURE CHART
TASK-LEVEL DESIGN MODEL
3.4
EVALUATE STRUCTURE CHART
3.3
DERIVE STRUCTURE CHART
STRUCTURE CHART
REFINED STRUCTURE CHART
DATA DICTIONARY ERD'S
DATA DICTIONARY + PROCESS SPECS
3.5
DESIGN MODULES
PHYSICAL CONSTRAINTS
MODULE DESCRIPTIONS
3.6
DESIGN DATABASE
DATABASE DESIGN
3.7
PACKAGE DESIGN
OPERATIONAL RESTRICTIONS
DESIGN SPECIFICATION

Figure 6.2

6.2 COMPONENTS OF THE DESIGN ACTIVITY

Let us examine each of the subactivities of design.

6.2.1 Subactivity 3.1: Allocate Specification to Processors

The first design subactivity is to allocate portions of the structured specification to individual processors; this is shown in Figure 6.3.

It should be emphasized that this subactivity was generally not relevant for most systems throughout the 1960s and 1970s; it was a foregone conclusion in most projects that all of the automated portion of the system would operate inside one CPU. Although this is still true in many cases, the designer generally has a number of options available: mainframes, minicomputers, and microcomputers in a variety of local and distributed configurations. Thus the first step of design is to deal with this level of system architecture.

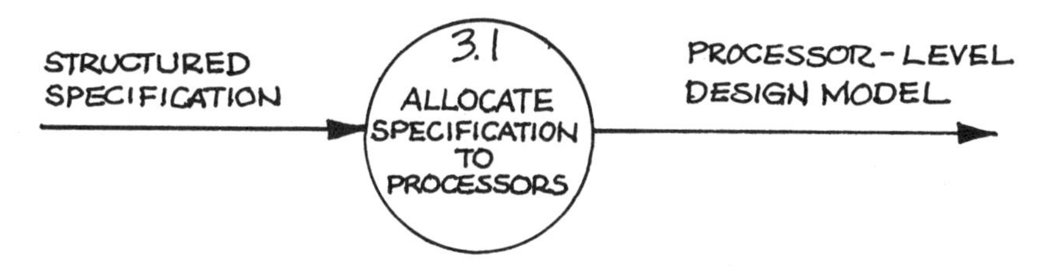

Figure 6.3

Hardware selection is an arduous, time-consuming, and often frustrating business, and I make no attempt in this book to tell you how to do it or even what detailed steps need to be followed. Many of the decisions about hardware selection will be made by someone other than the designer or even the project manager—decisions that are not already imposed on the project by the user may well be made by the operations department or some specialized group outside the project team.

It should be emphasized that subactivity 3.1 is essentially a *mapping* activity, mapping portions of the behavioral model produced in the analysis phase (processes and data stores) onto selected processors. This is illustrated in Figure 6.4(a). When viewed in this fashion, the specification can help the designer make an intelligent allocation of the behavioral model to individual processors. In virtually all real-world situations, the communication of data *between* processors is slow and costly compared to the communication of data from one function to another within the same processor. Hence it makes sense to partition the behavioral model in such a way that interprocessor communication is minimized wherever possible. Similarly, it is usually very

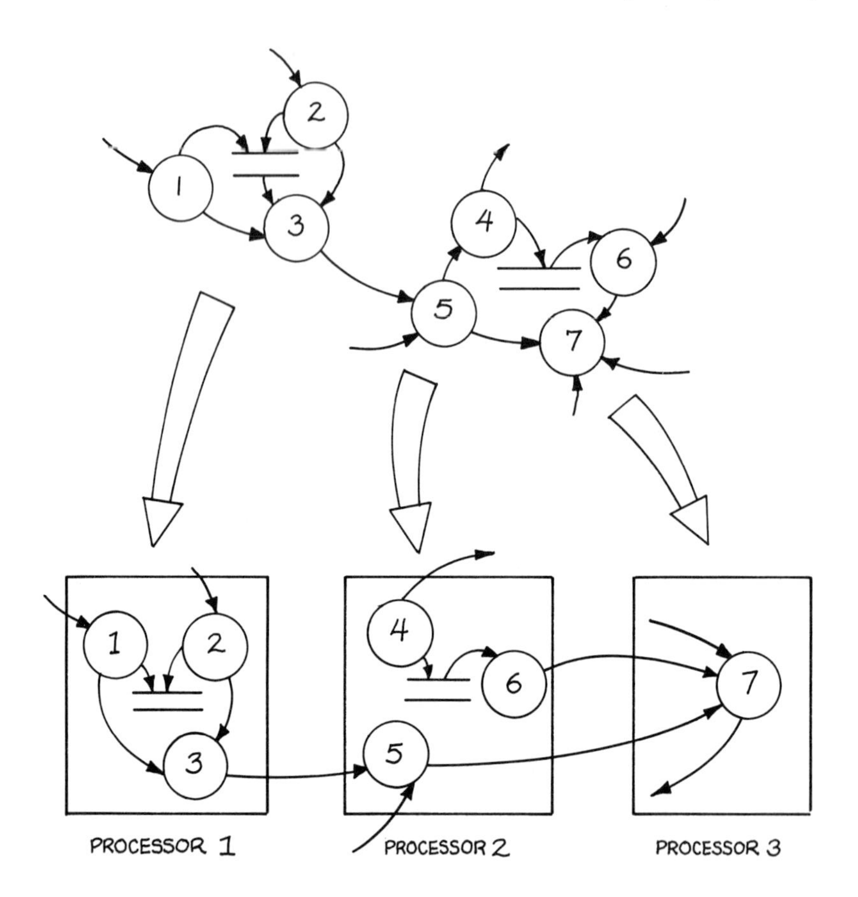

Figure 6.4(a) Allocating the specification to processors.

important, from the designer's point of view, to ensure that functions (bubbles) in the specification are not fragmented between processors any more than is absolutely necessary.

Note that the analysis activity may provide a number of constraints—or at least strong hints—about the system architecutre. During analysis, for example, the man-machine boundary is identified; the number of functions left inside the machine area will obviously influence the designer when he tries to determine the size of processor required. Also, the *nature* of the man-machine boundary may influence the designer's choice of hardware and systems architecture. If, for example, the systems analyst has defined the man-machine boundary as shown within the solid line in Figure 6.4(b), the designer is perhaps more likely to choose a multicomputer architecture; a man-machine boundary of the sort depicted within the dotted lines in the Figure 6.4(b) is perhaps more likely to lead to a single-machine architecture. In addition, the process of

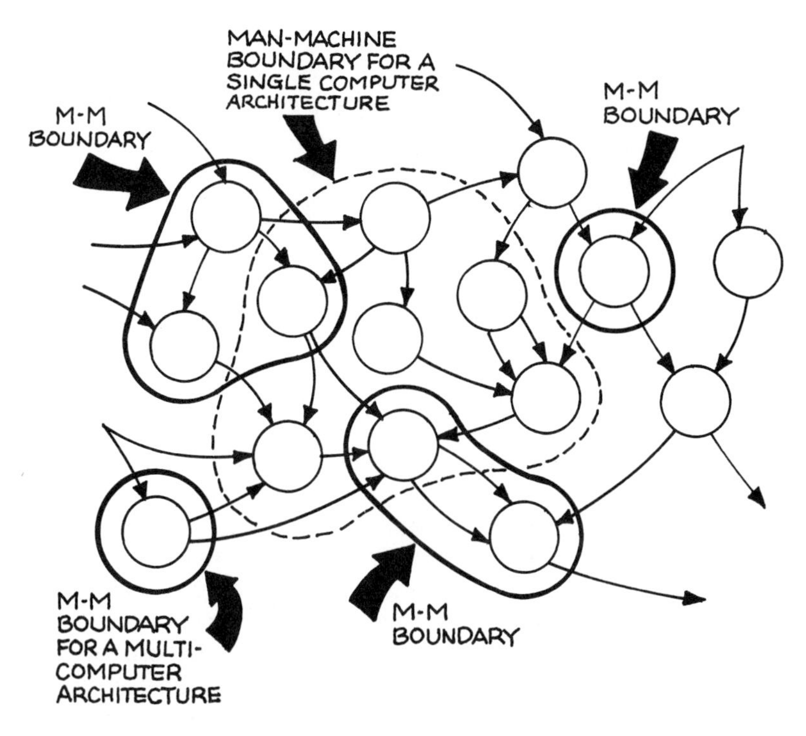

Figure 6.4(b) Man-machine boundaries determine computer architecture.

analysis produces a number of physical constraints, which have an enormous influence on the designer's choice of hardware; these constraints typically will include information about reliability, maintainability, and price considerations; environmental constraints; and political considerations, among others.

Data stores in the behavioral model must also be allocated, and the "mapping" approach illustrated in Figure 6.4(a) assumes that each processor has its own dedicated data storage facilities (disk drives, tape drives, etc.). However, it is possible that a data store might not be assigned to an individual processor but might be available to several different processors; for example, it might be implemented on a file server in a local area network.

Of course, choice of vendor, hardware configuration, and system architecture involves major decisions; consequently, it is likely that at least some part of subactivity 3.1 will begin before all of the analysis is complete. This is consistent with a philosophy that we have emphasized throughout this book: The activities in the project life cycle do not need to be carried out in a strictly linear fashion. Financial considerations, scheduling (deadline) considerations, and political considerations often make it necessary to begin working on design and implementation activities before the earlier activities are complete.

The result of subactivity 3.1 is a set of "miniature" specifications, each representing a portion of the overall specification that will operate inside each processor; I have referred to this as a "processor-level design model" in Figure 6.3. In addition to the set of DFDs, ERDs, and other diagrams for each processor, the designer should also, at this point, document the interprocessor interfaces—the communications protocol that will allow bubbles in one processor to communicate with bubbles in another.

6.2.2 Subactivity 3.2: Allocate Specification to Tasks

The second subactivity of design is similar to the first, but instead of mapping portions of the behavioral model onto processors, we now map portions of the model that have been assigned to an individual processor onto one or more "tasks" in that processor.

Again, this was typically an irrelevant activity through the 1960s and 1970s, since many large systems were implemented as a single monolithic program running within a single "partition," "job step," or "task." However, most vendor-supplied operating systems now provide a variety of sophisticated multitasking facilities so that several portions of the behavioral model (i.e., several bottom-level bubbles in the DFD) could be operating concurrently within the same CPU. Operating systems such as UNIX also facilitate the communication of data from one process to another, through "pipeline" connections that directly implement the data flow shown on the DFD.

The mapping activity is basically the same as that discussed in Section 6.2.2, and is illustrated again in Figure 6.5. Just as interprocessor communication tends to be slow and expensive, so it usually is true that intertask communication is relatively slow and expensive. Thus the designer should partition the processor-level model in such a way that intertask communication is minimized.

The output of this activity is a set of miniature models, one for each task within each processor; I have referred to this as a "task-level design model" in Figure 6.5.

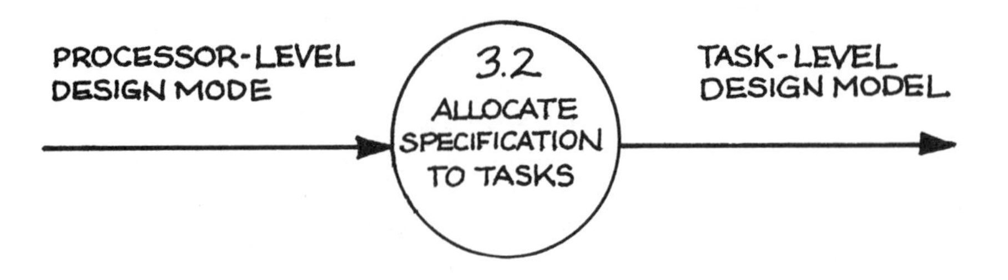

Figure 6.5

6.2.3 Subactivity 3.3: Derive Structure Chart

The primary purpose of this step is to produce a structure chart, or set of structure charts, for each of the task-level sets of data flow diagrams. Figure 6.6 illustrates the input and output to this activity.

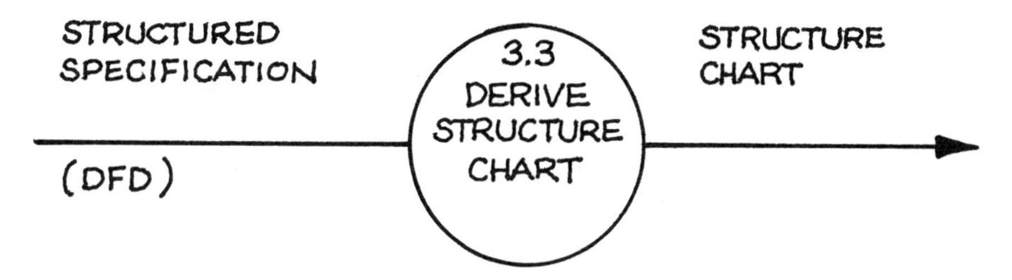

Figure 6.6 Derivation of structure chart.

In the context of our introductory figure and statement, the structure chart is defined as follows:

- *Structure chart:* a graphic representation of a hierarchy of modules, including a documentation of the *interfaces* between the modules

Normally, the designer will use the leveled data flow diagrams from the specification as a guideline for producing the structure charts; the levels of DFDs will correspond to the levels of structure charts. For example, if the top-level DFD has the form shown in Figure 6.7(a), it would guide the designer to develop the top-level structure chart shown in Figure 6.7(b). The next level of DFDs beneath the top level help the designer partition the high-level modules in his structure chart into appropriate lower-level modules; thus the lower-level DFDs shown in Figure 6.8(a) help the designer partition his structure chart as shown in Figure 6.8(b).

In many cases the designer needs little assistance in this process of developing a structure chart from the leveled DFDs; however, there are many cases in which the DFDs are far more complex than the simple linear DFDs

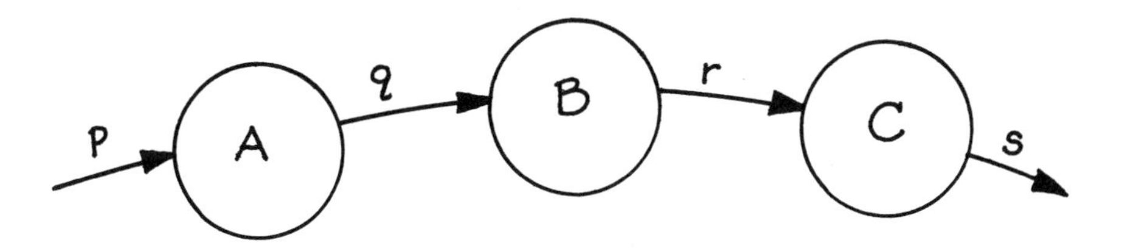

Figure 6.7(a) Top-level DFD.

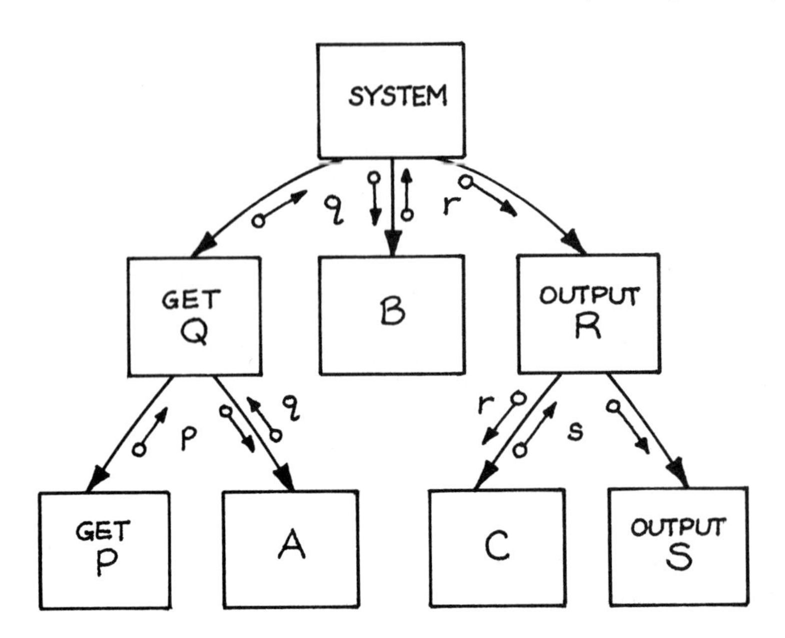

Figure 6.7(b) Top-level structure chart.

shown here. Formal design strategies such as *transform-centered design* or *transaction-centered design* are often necessary to help the designer create his initial structure chart. The concept of design strategies was discussed in Chapter 2; detailed treatment can be found in books by Page-Jones (1980) and Yourdon and Constantine (1979).

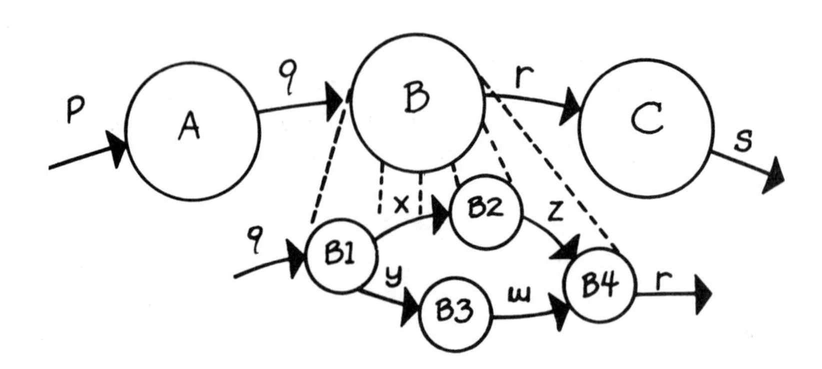

Figure 6.8(a) Detailed DFD.

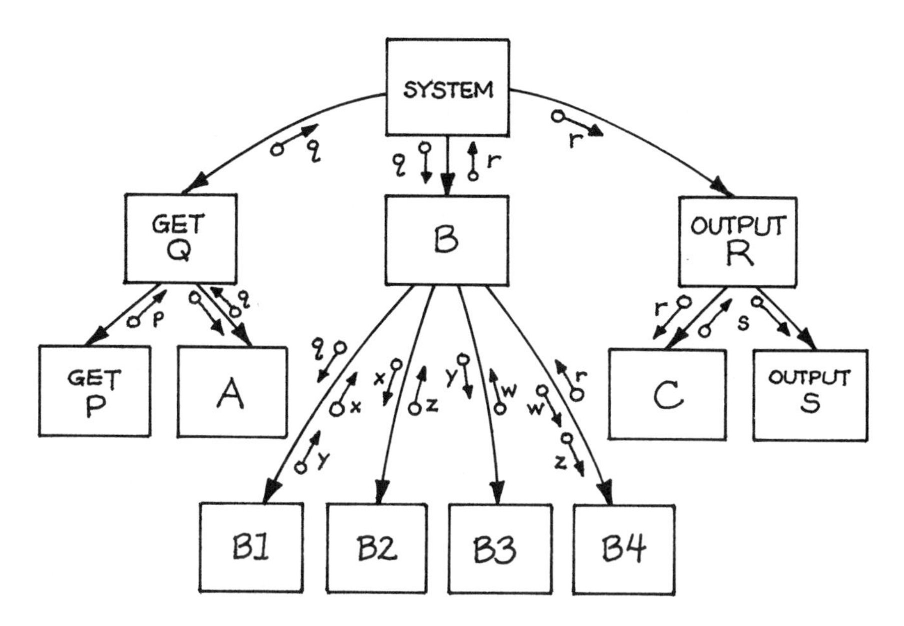

Figure 6.8(b) Detailed structure chart.

6.2.4 Subactivity 3.4: Evaluate Structure Chart

The fourth step in design is the evaluation of the set of structure charts produced in the previous step. Figure 6.9 depicts this activity and shows inputs and outputs.

If the data flow diagrams in the structured specification were derived properly and if the designer used a strategy such as transform-centered design in subactivity 3.3, the structure charts that are input to subactivity 3.4 should be a reasonably good tool with which to model a good design. However, it is possible that the structure chart(s) may have to be rejected, and subactivity 3.3 may have to be repeated.

In most cases the structure chart is acceptable but requires some refinements. Refinement should be based on such structured design concepts as coupling, cohesion, span of control, scope of effect or control, and module size.[1] To meet these criteria, the designer might move a module to a different place in the hierarchy, partition some modules into smaller modules, or combine accidentally separated parts of a function to constitute one module.

[1]If you are unfamiliar with these structured design concepts, see Yourdon and Constantine (1979).

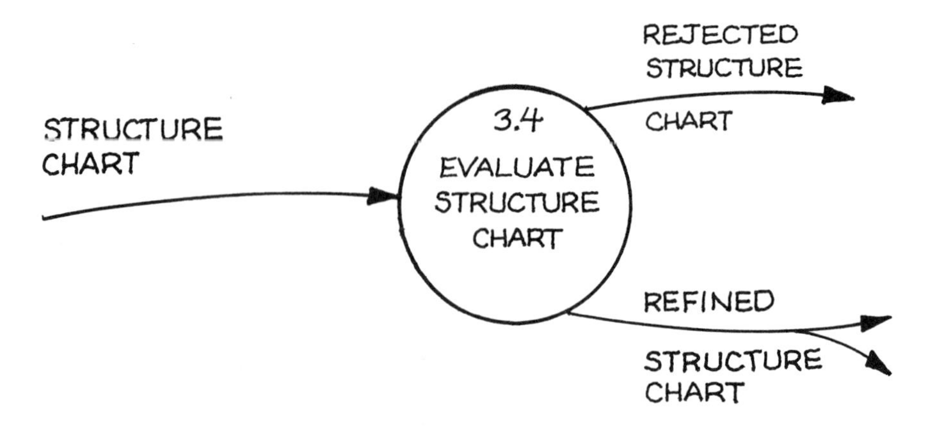

Figure 6.9 Structure chart evaluation.

As we can see from the diagram in Figure 6.9, the two possible outputs from subactivity 3.4 are either a rejected structured chart or a refined structure chart. These are defined as follows:

- *Rejected structure chart:* a structure chart that has been rejected because it depicts a design that has severe coupling problems, cohesion problems, etc.
- *Refined structure chart:* a structure chart that has been modified and revised to correct design problems, such as minor coupling problems, cohesion problems, span-of-control problems, and the like

It is important to note that this process of evaluating and refining the initial structure chart usually requires considerably more time and effort than the initial creation of the structure chart in subactivity 3.3. If the structured specification is a good one and if proper design strategies are used, it is unlikely that the initial structure chart from subactivity 3.3 will have to be rejected; however, you should not be surprised if the project team spends five times longer refining the structure chart than creating it initially.

6.2.5 Subactivity 3.5: Design Modules

The purpose of subactivity 3.5 is to carry out the detailed design of each module in the system. As we can see in Figure 6.10, there are three inputs to this process: process specifications, the data dictionary, and the refined structrure chart produced by subactivity 3.4. The output of subactivity 3.5 is a set of module descriptions, which are defined as follows:

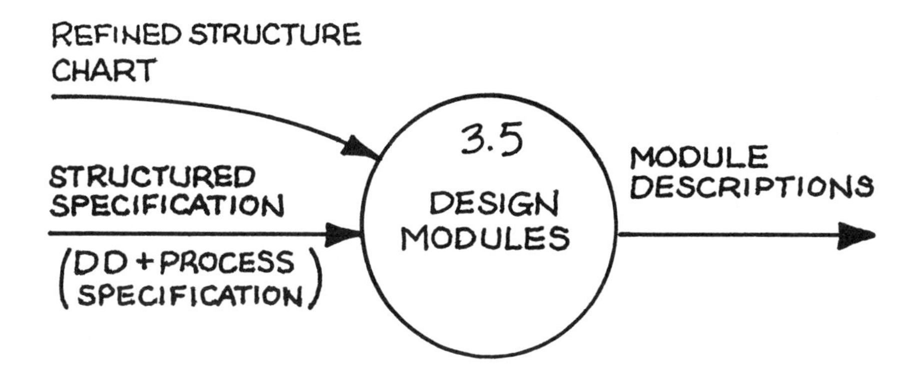

Figure 6.10 Designing modules.

- *Module descriptions:* detailed description of the procedural logic of a module, expressed in pseudocode, decision tables, Nassi-Shneiderman diagrams, or even (ugh) flowcharts

Assuming that the rest of the project is being conducted using structured design, structured programming, and so on, it would be common to see the module descriptions expressed in a format that encourages structured logic; *pseudocode* and *Nassi-Shneiderman diagrams* are among the more common tools for doing this.

In some cases subactivity 3.5 may turn out to be relatively trivial. This will be true, for example, if the procedural logic for the module is trivial or if the process specifications were written in a highly computer-oriented form of structured English (such as something that looked much like PASCAL) and the target language is highly structured (e.g., Ada, PL/I, or PASCAL).

In most cases, though, there is a fair amount of work to be done in subactivity 3.5. The process specifications, for example, might consist of a graph; translating that graph into algorithmic logic will require some effort on the part of the designer. Further, if the process specifications are written in a compact, high-level form, the designer may have to translate them into a more primitive language, like COBOL or FORTRAN. Thus a single process specification like "Find entry in transaction table" might be translated into a dozen or more pseudocode statements.

Note also that subactivity 3.5 also uses the data dictionary and the refined structure charts as inputs. The reasons for this are twofold: First, the refined structure chart will show end-of-file flags, switches, and other control parameters that were not present in the specification but will be an integral part of the detailed procedural logic. Second, the structure of the procedural logic will depend to some extent on the structure of the modules' input data elements and output data elements—hence the need for the data dictionary.

This last point is an important one. As we saw in Section 2.6, some of the major design strategies are based on data *structure* rather than data *flow*. Indeed, this is where the Jackson and Warnier-Orr techniques are usually best applied: to the design of the innards of individual modules. For more details on these strategies, consult the books by Jackson (1975), Warnier (1976), and Orr (1977, 1981) in the bibliography.

I should point out that it is not necessary to develop a detailed module design for every module in the structure chart. There will be some modules—particularly at the bottom of the structure chart—whose implementation is so simple and so obvious that the project manager may decide that it is not cost-effective to write pseudocode. However, this is a decision that the project manager should make, not the technician. As we will discuss at the end of this chapter, one of the common problems in the design activity is the technician's failure to do *enough* detailed design.

One reason for carrying out detailed design for the majority of modules in the structure chart is to reveal essential information about the quality of the structure chart itself. The structure chart may look very elegant, but the designers may find that it is simply impossible to pseudocode any of the modules! The ultimate test of the quality of the structure chart will come when *real* code is written and tested, but inability to produce decent pseudocode for the modules in the structure chart is an obvious sign that there is something fundamentally wrong with the design.

6.2.6 Subactivity 3.6: Design Database

The major purpose of subactivity 3.6 is to design the physical database, based on the *logical* information in the data dictionary and entity relationship diagram components of the structured specification, together with the physical constraints. This is shown in Figure 6.11.

Typical questions that are raised in performing subactivity 3.6 are, Which database package shall we use? (IMS? IDMS? TOTAL? ADABAS?) Which access method shall we use? (VSAM? ISAM? BDAM?) Shall we develop our

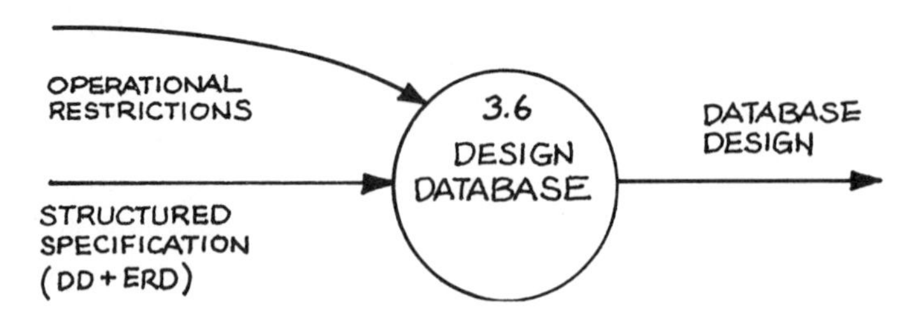

Figure 6.11 Design of a database.

own access method? Is a relational database practical for this application? What kind of record structure shall we develop? What kind of buffering and blocking, if any, shall we use? What kind of backup facilities and audit-trail facilities need to be built in?

The output of all this is a database design, defined as follows:

- *Database design:* document describing the *physical* design of the database, documented in whatever format is appropriate for the nature of the design

As we have seen, there is a simple universal notation that we can use to document the logical structure of the data elements and data stores in the system: the data dictionary and the entity relationship diagrams. But the documentation for the *physical* database will almost certainly reflect the terminology and the general nature of the design approach; that is, one would expect different documentation for a sequential tape file database than for an IMA database.

6.2.7 Subactivity 3.7: Package Design

At first glance, subactivity 3.7 seems clerical. It gathers together the various products of the other design subactivities and produces a final document, the design specification. This is shown in Figure 6.12.

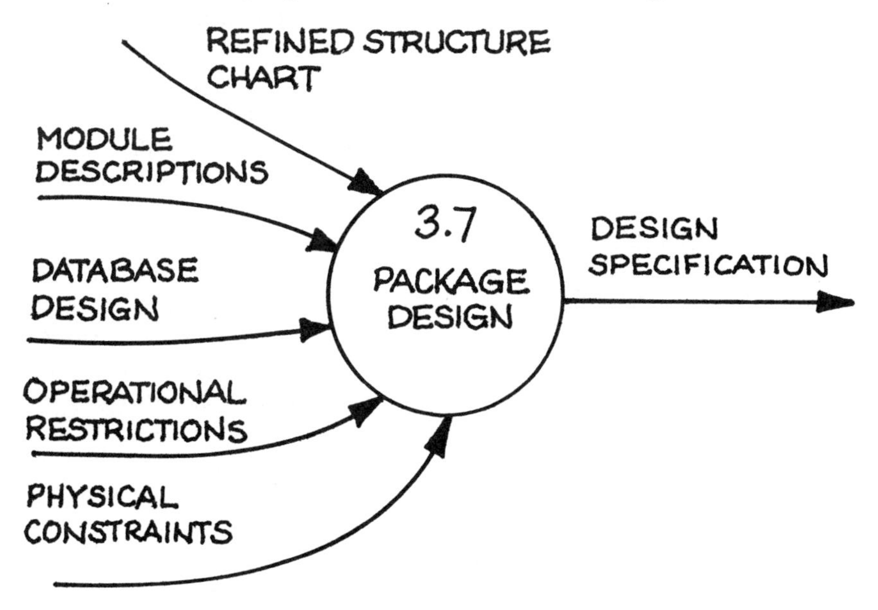

Figure 6.12 Packaging of a design.

Subactivity 3.7 in one sense resembles subactivity 2.8, in which various documents were packaged together to produce the structured specification. However, there is an additional task to be done in subactivity 3.7: The design must consider the impact of the entire design on the physical environment (the processors and tasks) in which it must run. The typical questions that are asked here are, Is there enough memory for all of the modules? Will overlays be needed? Do we need intermediate files for checkpoint and restart? In addition to considering these issues, one could argue that the *form* of the modules is a packaging decision (in COBOL, for example, one could use either the PERFORM statement or the CALL statement to transfer control from one module to another). The choice need not be made during earlier stages of design, but it must be made before subactivity 3.7 can be completed.

The ultimate packaging decisions consist of hardware selection and an overall system configuration; these were accomplished in subactivities 3.1 and 3.2. Thus the work carried out in subactivity 3.7 represents the last low-level decision about how best to allocate portions of the specification to implementation technology.

6.3 TOP-DOWN CHOICES FOR DESIGN

As we have noted in earlier chapters, the project manager can elect various degrees of radical or conservative implementation; obviously, this is true for the design activity, too. For example, the ultraradical approach to design would involve carrying subactivities 3.1 through 3.7 for only the top two or three levels of the structure chart and only the top two or three levels of the data. With only that much of the design accomplished, one could proceed to the implementation activity. There are dangers associated with this approach, as we discussed in Chapter 3, and in most cases I would not recommend it.

The moderately radical approach is less dangerous. In this case we would imagine the design effort continuing to approximately halfway to the bottom of the system. For most medium-sized systems this would mean roughly five to ten levels of modules in the structure chart and would involve design of the database to the field level. This should provide the designer with enough information to know whether he has a solid design—that is, he should feel reasonably confident about coding and testing the top of the structure if he has designed ten levels deep.

The moderately conservative approach involves performing the design to within two or three levels of the bottom; indeed, the designer will probably announce that he can "see" the bottom of the design, although he will not actually have recorded it. Of course, this should provide enough substance to the design that the designer should have no hesitation about commencing the implementation. Indeed, the major argument against the moderately conservative approach is that it may take too much time—time during which there is no tangible evidence of progress.

Finally, there is the ultraconservative approach. Obviously, this would entail a *complete* design. This presumably would take even more time to finish, and it runs the risk that upon seeing the first few working versions of the system, the user may change the requirements of the bottom of the system, thus wasting all of the bottom-level work that the designer has so carefully done.

6.4 PROBLEMS WITH DESIGN

The design activities described in this chapter don't always go according to plan. Let us review the most common problems to watch out for.

First is an especially common complaint on small or even medium-sized projects: *There is not enough time to do "proper" design.* The designer (who is usually also the implementer) is so eager to begin programming that he is reluctant to bother with the tedious detail of actually writing down the design. "Jeez," he'll say, "by the time I complete all of this structure chart stuff, I could have the whole thing coded." Any project manager who doesn't have several good responses to this comment ought to have his head examined.

A subset of this problem arises because many designers are unwilling to write detailed pseudocode for their modules and will argue that subactivity 3.5 is really part of implementation and *not* something that they as senior-level technicians should have to do. If it is true that the application is relatively trivial, the process specifications were written in a PASCAL-like form of structured English, and the system will eventaully be implemented in PASCAL or Ada, the designer might have a valid point. But if the application is complex, the process specifications are written in a language resembling unstructured Urdu, and the system will eventually be implemented in the original FORTRAN, the act of pseudocoding the modules is exceedingly important and is indeed a part of the design activity. It may offend the designer's political sensibilities (after all, what will happen if a grade 17 senior software systems analyst is seen writing something suspiciously close to COBOL?), but it is nevertheless an important aspect of design.

Next on the list are *fuzzy structure charts,* which are not accurate, technical models. This problem is common among designers who use HIPO as their form of design documentation. It's not that anything is wrong with HIPO, but the way people *use* HIPO is such that there is *not* a one-to-one correspondence between boxes on the HIPO diagram and actual modules in the coded system. It thus becomes difficult to evaluate the technical merits of the design and virtually impossible for the maintenance programmer to establish any correspondence between the design documents and the code he is trying to debug.

Another common problem occurs in connection with the *random genera-*

tion of structure charts. You'll recall that subactivity 3.3 involves the translation of data flow diagrams into structure charts—and you'll recall that I suggested the use of transform-centered design or transaction-centered design as orderly methods for accomplishing this translation. Unfortunately, many designers, unaware of these design methods, attempt to create the structure charts through their own intuitive approach. There's nothing wrong with intuition, or black magic for that matter, but you should watch out for one common danger: The resulting structure chart may not match the data flow diagrams from which it was supposedly derived.

There seems to be some confusion between "guidelines" and "commandments, " especially when structured design is being used for the first time on a software development project. The resulting problem is endless *arguments* about coupling, cohesion, and other aspects of structured design. You should be particularly careful about this as a project manager: Your concern should be with design techniques that work rather than a blind adherence to a particular set of words in a textbook (including this one!). A certain amount of discussion and debate on the technical merits of a design is extremely healthy, but at a certain point it's up to you as a mangaer to cut off the intellectual nit-picking and instruct your designers and programmers to get down to work.

An obvious problem occurs when the *design drifts away from the structured specification.* The fact that drifting occurs has nothing to do with the use of structured analysis, structured design, or any other particular discipline. It is an especially serious problem if the analysis is done by one person (or group) and the design is performed by another. The most effective solution is a formal review, attended by people other than the designers (for example, users, systems analysts, quality assurance personnel, and others), to ensure the integrity of the design. To avoid this problem, it may be a good idea to develop a matrix that shows where each piece of the system specification is addressed within the design.

6.5 SUMMARY

Fortunately, many of the concepts of structured design discussed in this chapter have appeared in the literature since the mid-1970s. Thus most of your technical staff should be aware of the concepts; your younger employees have probably been introduced to structured design in their university courses. Thus the design activity should not present you with any major problems other than the normal problems of selecting hardware, getting it delivered on time, and so on.

There is one sociological problem that should be mentioned again: Senior-level people are accustomed to carrying out high-level conceptual tasks, whereas junior-level people can be expected to carry out the more detailed, mechanical aspects of EDP systems development. This point was mentioned in our discussion of systems analysis; it is equally true in the area of design. A designer with ten years' experience may resent being asked to write detailed pseudocode for low-level modules. It makes sense to give that job to junior-level staff members who can also produce the final code from their own design (thus giving them a sense of accomplishment, rather than making them feel like clerical-level coders).

The close relationship between the pseudocode and the final COBOL or PASCAL coding, together with the fact that both are likely to be written by the same person, leads us to another theme that has been repeated throughout the book: parallel progress through several of the major activities of systems development. If a junior-level designer writes the detailed pseudocode for a module and if he will eventually write the final PASCAL version, why not have him write the "real" code right away, while it is fresh in his mind?

7

Activity 4: Implementation

7.1 INTRODUCTION

At this point in the project life cycle, we are finally ready to produce some code. During the implementation activity, code is written and a major part of the testing is carried out. Inputs and outputs contained within the implementation activity, as well as the overall context of the implementation activity, are shown in Figure 7.1.

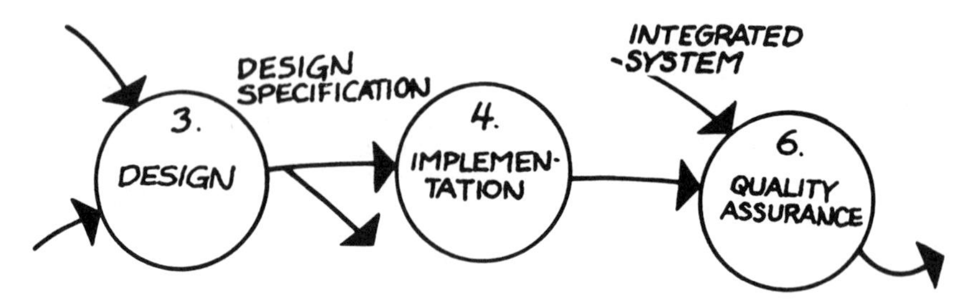

Figure 7.1 Context of implementation.

7.2 THE STEPS OF IMPLEMENTATION

Figure 7.2 gives an overview of implementation. We will examine each of the three distinct subactivities.

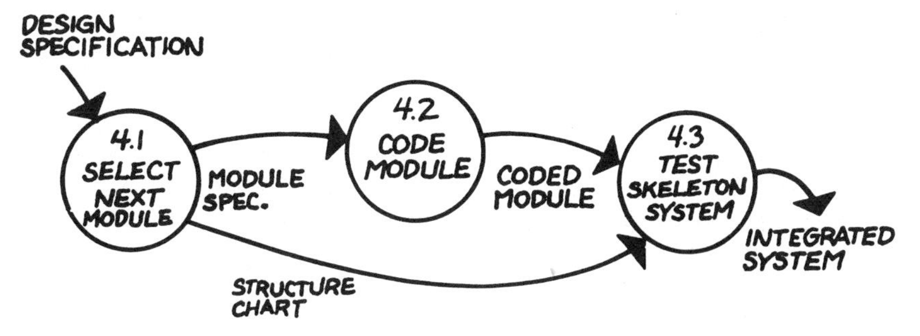

Figure 7.2 Overview of the implementation activity.

7.2.1 Subactivity 4.1: Select Next Module

The first step in the implementation activity is to decide the sequence in which modules will be implemented and, in particular, which module will be implemented next; this is shown in Figure 7.3. Subactivity 4.1 is simply a decision-making process, based on two fundamental assumptions. The first is that the system will be coded, integrated, and tested *incrementally;* that is, one new module at a time will be added to an existing skeleton system, and the resulting system will be tested. This is an extremely important principle: With incremental development, the testing process is more orderly, and the *debugging* process (which is related to, but nevertheless distinct from, testing) is considerably simpler. Second, it is assumed that the system will be developed *from the top down.* Modules at the top of the hierarchy will be coded and integrated before modules at the bottom of the hierarchy.

Even though it is assumed that the system will be built incrementally and from the top down, there is still the question of which module to implement next. For example, the user may be more interested in some features

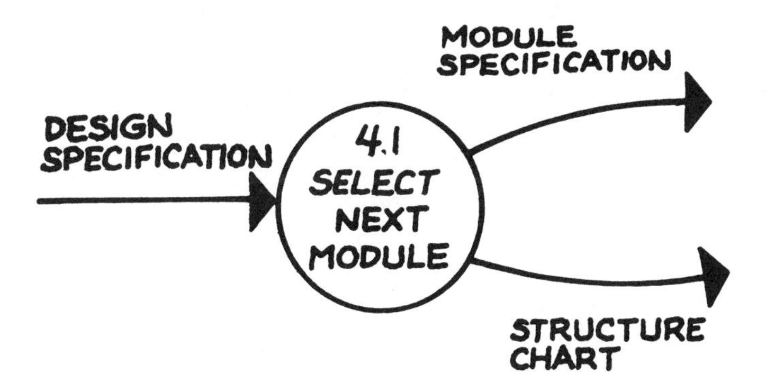

Figure 7.3 Module selection.

of the system than in others; that may dictate the sequence in which the modules are implemented. Or in a multiperson project, the manager will discover that some programmers are faster than others, and *that* may dictate the sequence of development.

Typically, there are three distinct stages of top-down development, as illustrated in Figure 7.4. The first version, or series of versions, is aimed at building the umbrella-shaped structure shown in the figure. That is, it creates a system that is able to obtain input, although very little formatting, error-checking, or editing will take place, and to produce some output, although it may be crudely formatted. Typically, no computations will take place; indeed, in a top-down, on-line system, the first version may do nothing more than echo the characters that the user types on his terminal.

The next several versions typically involve "horizontal slices" of the system, in which actual modules replace "stubs" or place holders used to

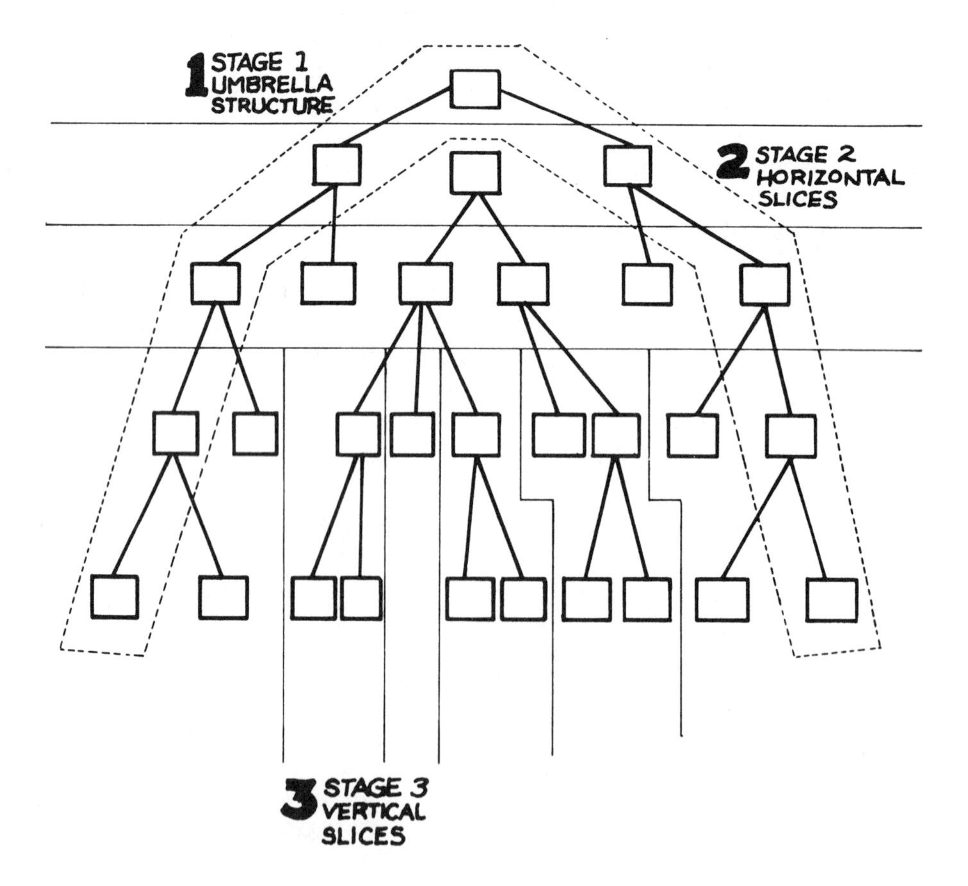

Figure 7.4 Top-down development stages.

emulate their presence in previous versions. That is, all of the modules at level 2 will be implemented, one at a time, then all of the modules at level 3, and so forth. Toward the end of the development, when the developers are near the bottom of the system, the strategy will probably switch to vertical slices of the system. That is, the developers will pick one of the modules and implement *all* of its subordinates to the very bottom level; then another module will be selected and its subordinates implemented, and so on.

When the selection process has been completed, the module specification is sent to subactivity 4.2 for coding and the structure chart is sent to subactivity 4.3 for integration.

7.2.2 Subactivity 4.2: Code Module

Code gets written during subactivity 4.2. The input to subactivity 4.2 is the module specification produced by subactivity 4.1—it is an indication of which module code, together with the design specification for that module produced during design. The output is a coded module, which, under normal circumstances, will be sent to subactivity 4.3 for integration. This is shown in Figure 7.5.

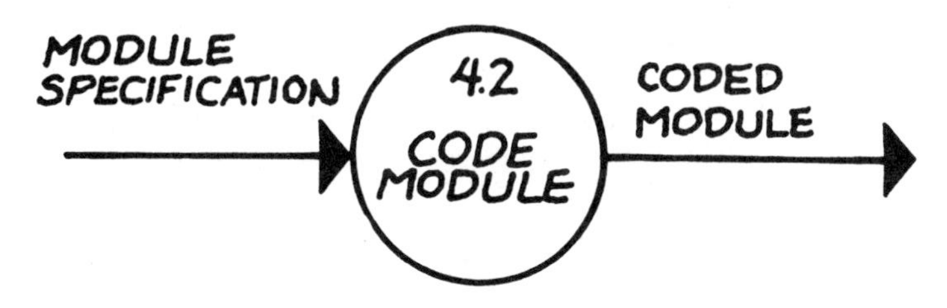

Figure 7.5 Module coding.

In some cases it will be appropriate to conduct some stand-alone "unit" testing before sending the coded module on to the integration step. If good test facilities are available—for example, if the developer has access to time-sharing terminals or a workstation with local testing and debugging facilities—it may save time and energy to eliminate the trivial logic errors inside the module before integrating it into the system.

Walkthroughs are also appropriate at this point, not only to ensure the correctness of the code but also as a form of quality assurance.

7.2.3 Subactivity 4.3: Test Skeleton System

Subactivity 4.3 simultaneously performs module testing and system integration. The coded module produced by subactivity 4.3 is combined with the partial system that already exists, and the resulting system is then tested. This is shown in Figure 7.6.

As previously mentioned, in almost every case the coded module replaces a stub that cxistcd bcforc. For cxamplc, if thc ncwly coded module has the function of computing an employee's net salary, it may replace a stub that happily paid all employees (regardless of rank or seniority) a salary of $100.

Note that one of the inputs to subactivity 4.3 is the structure chart produced by the design activity. This is needed as input because the developer must learn from the structure chart what kind of stubs, togther with the newly coded module, are required to produce the new skeleton system.

At this point, I am assuming that the test data have been generated and that the testing process has been conducted by the programmers who wrote the code in subactivity 4.2.[1] It is appropriate and highly advisable to have informal walkthroughs and even formal reviews to ensure that the test data are reasonably comprehensive and that the testing process is carried out properly in subactivity 4.3. Nevertheless, subactivity 4.3 represents "internal" testing within the development organization. "External" testing, in which the users are involved, is discussed in Chapters 8 and 9.

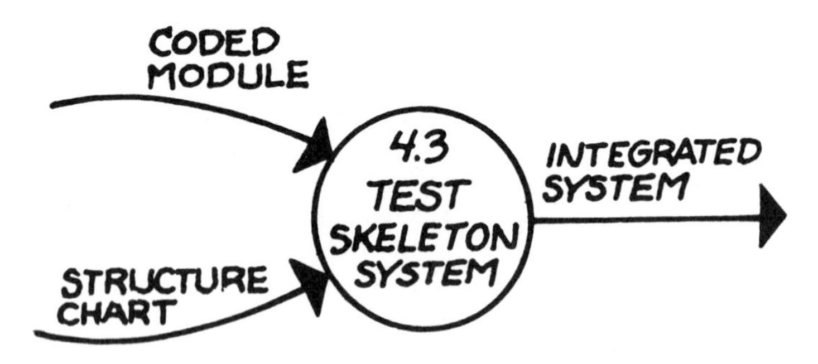

Figure 7.6 Test skeleton system.

7.3 TOP-DOWN CHOICES FOR IMPLEMENTATION

As we have seen in the earlier chapters, all of the activities in the project life cycle can be approached in either a conservative or a radical fashion. This is true of implementation, although you should keep in mind that the definition of the final product of implementation, the integrated system, is itself subject to top-down interpretation. That is, the developers may make a complete pass through the implementation activity to develop a system whose design is not yet complete.

[1]Note, though, that in some organizations someone other than the code author may be given the job of performing an independent testing function.

Within the implementation activity itself, the developers have the option of pursuing an ultraradical approach, a moderately radical approach, a moderately conservative approach, or an ultraconservative approach. The ultraradical approach would involve coding and integrating in parallel, one level at a time. Programmer A might be coding one module at level 2 of the hierarchy while programmer B is simultaneously integrating another level 2 module into the system.

A moderately radical approach would involve coding two or three levels ahead of the integration work being carried out in subactivity 4.3. One reason for suggesting this approach is that the act of coding often uncovers subtle design errors (not previously apparent to the designers), which are easier to correct while coding than they would be after coding.

Similarly, the moderately conservative approach involves coding to within two or three levels of the bottom of the hierarchy before any integration takes place. And the ultraconservative approach would involve coding all the modules before any integration had taken place. It is hard to imagine why the designers would want to pursue either of these conservative approaches, unless forced by lack of available computer resources for testing or other external pressures.

7.4 PROBLEMS WITH THE IMPLEMENTATION ACTIVITY

Many of the problems I have observed at this stage of a typical development project are properly labeled *management* problems; hence they should be of particular interest to readers of this text. The four most common problems are described.

First there may not be sufficient hardware. The top-down approach described in this chapter clearly requires computer hardware—and it usually requires hardware *earlier* in the project than was true with conventional, bottom-up projects. How much earlier you'll need the hardware depends, to a large extent, on whether you adopt a radical or conservative approach to the development of your system. But if you think that you can plug in the hardware the day before your system is scheduled to go into "live" operation, you're going to be in for a lot of surprises!

Second, test time may be inadquate. Not only do the developers need access to computer hardware for their testing, but they need it *frequently*. If they are forced to work in a batch development environment, with overnight turnaround for compiling and testing, they will be sorely tempted to avoid the incremental testing approach advocated in this chapter. They will be tempted instead to plug in a dozen modules at a time, cross their fingers, and hope desperately that it all works. To help enforce the discipline of incremental top-down development, the developers typically need several short bursts of computer time; hence, workstations, time-sharing terminals, or a remote job entry system with 5-minute turnaround is ideal.

Third, discipline problems may appear in multiteam projects. If you have a large project, with 30 or 40 developers, you will probably organize them into groups of five or six people each. Unfortunately, each group has a natural tendency to isolate itself from the other groups and then to complain about communication problems. "It's a real hassle," each group complains, "trying to get those guys in the frammis subsystem group to tell us how we're supposed to interface with them."

The result is often that each group does top-down program development but the system as a whole runs the danger of being developed from the bottom up. However, one could argue that this is not a problem with the top-down approach, but rather one of its benefits: After all, communication problems like this will exist in any project, whether it is being developed bottom up, top down, or inside out. The earlier these communication problems can be forced into the open and resolved, the better off everyone will be. And in a top-down project, the typical experience is that the first skeleton version of the system is exremely difficult to build, simply because of fuzzy or misunderstood interfaces. Subsequent versions tend to get easier and easier, and if you think about it, that's precisely what you want: to find the difficult bugs first, so that by the end of the project you're left with nothing but trivial bugs.

A final possible problem is deviation from the design specification. This is always a problem; as we discussed at the end of Chapter 6, deviations are usually more serious when different people perform the various activities in the project. That is, if the programming is performed by people other than the ones who did the design, there is more chance that the code will do something more, less, or different from the design specifications. Indeed, one common example of this is the programmer who attempts to fill gaps in the specification by mind-reading. The solution, obviously, is a careful review of the code by people who are familiar with the design document as well as the principles of good coding.

7.5 SUMMARY

The idea of combining coding and testing is probably quite a departure from the classical approach to project management—yet it is a natural result of the incremental, top-down testing concept discussed in this chapter.

You should have few, if any, problems with the task of programming: After all, programming is something that virtually *everyone* feels he understands. If systems analysis and design have been done properly, there should be little or no difficulty in producing the final code—indeed, it may become almost mechanical.[2] But while programming may be simple, the work of integrating coded modules into a skeleton system is not always so straightforward, and many of the problems mentioned in Section 7.4 may plague your project. However, I have a rather positive attitude toward all of this: I favor an implementation strategy that forces integration problems and interface problems into the open as soon as possible; this is precisely what happens with the top-down approach—as compared to the bottom-up approach, where the most serious problems are usually the last to be found, and time to resolve them is scarce.

Note, by the way, that the testing and integration we're talking about here is internal testing—testing done by the programmers and designers themselves. Our project life cycle assumes that there is a separate activity for developing formal acceptance tests, an activity that has probably been taking place in parallel with the activities of design and implementation and will probably be conducted by users, systems analysts, and other technicians who are not directly involved with the implementation of this system. The details of this external testing are discussed in Chapter 8.

[2]Ultimately, of course, everyone (except possibly the programmers, who worry that they will be put out of work) hopes that generation of code will become mechanical. At that point, all the emphasis will be on systems analysis and design.

8

Activity 5: Acceptance Test Generation

8.1 INTRODUCTION

The fifth major activity of the structured systems life cycle is the generation of test data that will be used for acceptance testing. Figure 8.1 shows the context of the acceptance test generation activity. As you can see from the figure, the sole input to the acceptance test generation activity is the structured specification produced by the analysis activity; its output is a set of acceptance tests that will be used by the quality assurance activity.

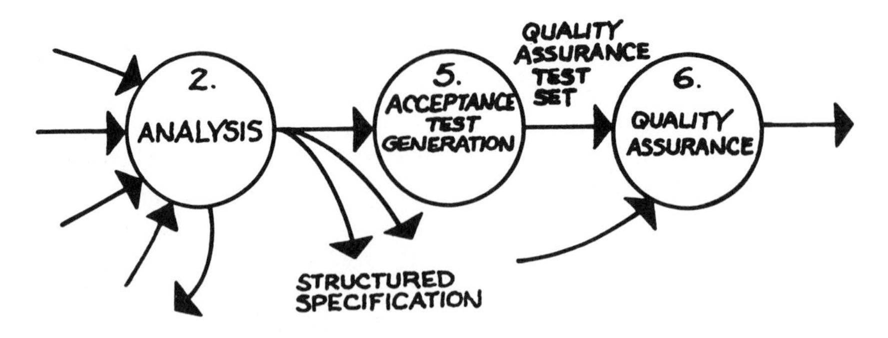

Figure 8.1 Context of acceptance test generation.

8.2 STEPS OF ACCEPTANCE TEST GENERATION

Figure 8.2 gives an overview of the acceptance test generation activity. Although each of the five subactivities will be discussed in detail, they may be summarized as the generation of test plans; the preparation of performance tests, normal path tests, and error path tests; and the packaging of the tests.

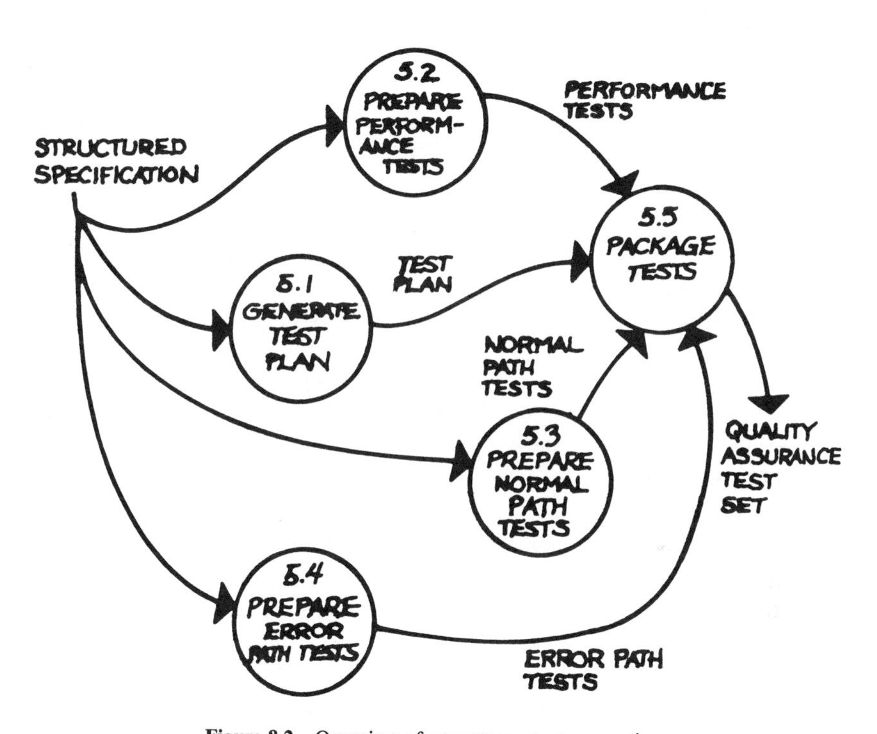

Figure 8.2 Overview of acceptance test generation.

8.2.1 Subactivity 5.1: Generate Test Plan

Subactivity 5.1 is probably the most important step of acceptance test generation, as it involves *development of a test plan;* this is shown in Figure 8.3. Once developed, the set of test plans is used by all subsequent activities.

The nature of the test plan will obviously vary from project to project; but without some kind of organized plan, there is little hope of building a well-tested system. The test plan should cover establishment of personnel to test the system, an outline of testing guidelines, criteria for determining test completion, and a statement of estimated time and material resources necessary

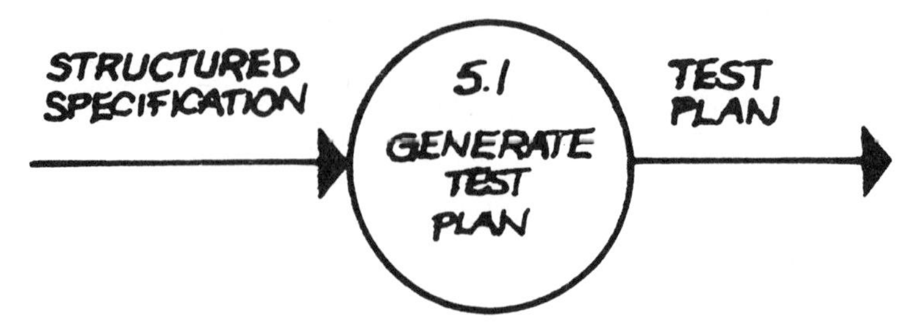

Figure 8.3 Generate test plan.

during the testing activity. The balance of this section deals with each of these four elements in turn.

The initial goal of the test plan should be *establishment of a person or group responsible for testing the system*. For obvious reasons, the person responsible should be someone other than the developer.[1] Indeed, the testing may well be done by people entirely outside the project team, in an effort to make the procedure as objective and thorough as possible. In some cases a formal testing group or quality assurance group within the EDP organization may be given the responsibility for testing the system; in other cases the user may even contract with an external organization (such as a software consulting firm) to do the testing. In any case, testing should be done by someone who understands that the process has the *express intention* of finding errors—that is, the test should be a thoroughly diabolical attempt to cause system failures.

Standards need to be developed for the construction of test cases, since this will affect subactivities 5.2, 5.3, and 5.4, and it is a goal of the test plan to *establish procedures and standards for testing*. There should also be standards for the documentation of test cases and results, naming conventions for sets (or files) of test data, and standards for storing and retrieving sets of test data. The reason for this emphasis on standards and procedures is simple: The sheer volume of test files, test cases, and test results is likely to be utterly overwhelming! During one project on which I recently worked, the developers produced roughly 200 modules and 10,000 lines of source code. To test the system properly, they required substantially more than 1000 separate test files.

Also essential to the success of the test plan for acceptance test generation is the setting of *criteria for completion for various forms of testing*. Clearly, the completion criteria should emphasize quality of testing rather than quantity—the fact that 117 hours of computer time have been consumed in the testing process is no guarantee that any significant testing was done! Some reasonable criteria for completion might be the following:

[1]Recall that developers were responsible for their own integration testing, which was discussed in Chapter 7 as part of the implementation activity.

- Every module in the system must have been invoked at least once, although not all possible combinations of modules may have been executed.
- Every program statement must have been executed at least once, although not necessarily with the full range of data elements.
- Every decision (e.g., IF-THEN-ELSE statement) must have been exercised, although not all combinations of decisions may have been exercised.
- A certain number of known bugs might be secretly seeded in the system, with the number of seeded bugs found during the testing process used to determine the effectiveness of the test data.
- Results of all test cases where the actual results differed from the expected results have been analyzed and appropriate action (e.g., revision of the code or of the test set) has been taken.

The final elements to be covered by the test plan are the *schedule, activities, and resources required during the testing effort.* This information is passed directly to subactivity 5.5 and eventually becomes input to the quality assurance activity, where testing is actually performed.

8.2.2 Subactivity 5.3: Prepare Performance Tests

The primary task of subactivity 5.2 involves coverting the time and volume requirements in the physical constraints document into specific tests to ensure that the constraints have actually been met. Figure 8.4 depicts input and output. Subactivity 5.2 is clearly essential for large on-line systems but may be unnecessary for small batch systems or for any kind of system in which the hardware is a tiny part of the overall cost of the system and is easily expandable (e.g., by adding more memory or upgrading to a larger model CPU).

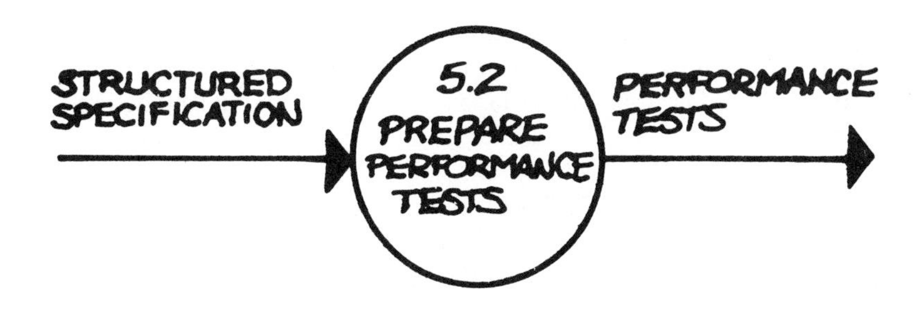

Figure 8.4 Prepare performance tests.

The output of subactivity 5.2 is a set of performance tests, which we have defined in data dictionary format as follows:

- *Performance test:* procedural description of a test of the performance of the system, complete with test data and expected results

In most cases, performance testing is concerned with *throughput, response time,* and *capacity*—especially capacity of the database. It may not be practical, or even possible, to create enough real input and enough of a load on the new system to verify that its performance is satisfactory; the testing group may have to resort to a variety of simulation techniques.

For example, initial systems analysis and design of the system may have included formal mathematical models or programmed simulation models, such as models developed in SIMSCRIPT or SIMULA. Once the system has been coded, *actual* performance data for small volumes of input can be supplied to the simulation model, and it should be possible, by extrapolating, to get a good feeling for how the system will perform with much larger volumes of input.

In other cases it may be possible to trick the system into operating as if it had a heavy volume simply by reducing the amount of available hardware; for example, instead of running the system of a 4-megabyte machine, run it on a 1-megabyte machine. In still other cases performance testing may be accomplished by a much simpler expedient: A small volume of test input can be replicated hundreds, or even thousands, of times to provide the appropriate high-volume test.

Naturally, these simulations and subterfuges are not guaranteed to produce perfect results. Indeed, unless the simulations are planned very carefully, they can give the test group misleading or false information about the actual performance of the system. For example, reducing the available memory on the computer from 4 megabytes to 1 megabyte may create thrashing or excessive swapping of programs—performance problems that would never be present in the actual system—and it may hide other performance bottlenecks that are important to find.

The preparation of performance tests clearly involves simulation, modeling, and capacity measurement—subjects that are beyond the scope of this book. But it should be apparent from this brief discussion that subactivity 5.2 is extremely important and should be given ample time and resources.

8.2.3 Subactivity 5.3: Prepare Normal Path Tests

This activity involves analyzing the structured specification in order to generate a set of test cases to verify functional conformity of the system—

that is, to confirm that the system does what it's supposed to do when given valid input data. This is shown in Figure 8.5.

The output of subactivity 5.3 is a set of normal path tests, defined as follows:

- *Normal path test:* procedural description of a test of the correct processing of valid data + test data + expected results

Simple as this may sound, subactivity 5.3 usually represents a massive amount of work, as test cases should be derived for *every* item in the data dictionary and for *every* composite data flow. However, it is well known to everyone in the systems development profession that generating every possible test case for every possible data element is impossible; instead, it makes sense to concentrate on areas where bugs tend to cluster, particularly the boundary

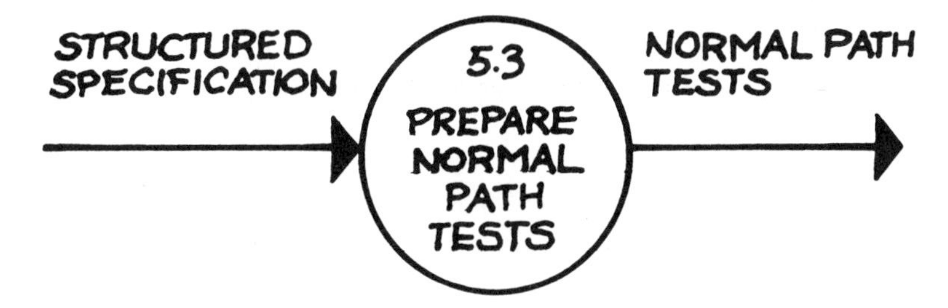

Figure 8.5 Prepare normal path tests.

situations.[2] For example, if the structured specification indicates that data element X is an input to the system and has legitimate values of any integer between 1 and 20, it would make sense to generate test cases of 1, 2, 19, and 20, with perhaps a few test cases in between.[3] In general, the test cases should concentrate on the following aspects of the system:

- *Input boundaries:* upper limits, lower limits, and other boundaries associated with inputs to the system. The purpose of these tests is to verify that the system will accept inputs that are within the limits described in the specification. Thus if the specification indicates that one of the system inputs is a character string between 1 and 40 characters in length, we

[2]For an excellent and very thorough discussion of the generation of proper test cases for software, see Myers (1979).

[3]Of course, it would also make sense to generate test cases of 0, -1, 21, 22, and 999999999999, not to mention 37QW45 and A#%&$@!. However, these constitute *error* test cases, which are included in subactivity 5.4.

want to ensure that it actually *will* accept a 1-character string, a 39-character string, and a 40-character string.

- *Output boundaries:* upper limits, lower limits, and other boundaries associated with the outputs produced by the system. For example, if a payroll system is supposed to be able to produce six-digit paychecks, it is obviously a good idea to construct test cases that will produce such paychecks.
- *Functional boundaries:* to ensure that the functions that are described in the process specifications actually perform properly, especially in the limiting cases. Suppose, for example, that a module TABLESORT has been specified to sort a table of items. Functional boundaries would be exercised by determining what happens when the table is empty, when the table has only one entry, when the entries in the table are already in sequence, when all the entries in the table are the same, or when the table is full.

8.2.4 Subactivity 5.4: Prepare Error Path Tests

The purpose of subactivity 5.4 is obvious, although many people would argue that it could be combined with subactivity 5.3. After all, the boundary between good data and bad data is often a fine one. A comparison of Figure 8.6 and Figure 8.5 shows little discernible difference.

However, there is one primary reason for separating the activities of normal path tests from those of error path tests: A different psychological makeup characterizes people who are good at devising and performing error tests. Almost anyone can think of normal cases to test; it requires a different personality to concoct error cases. Some of the error cases can be derived quite simply from the boundary cases discussed previously; after all, if we are dealing with a data element that can take on integer values between 1 and 20, it doesn't require too much ingenuity to generate test cases for -1, 0, 21, and 22. But it requires a slightly different personality to try a nonnumerical test case like A132PQ%)(&#)*Y!

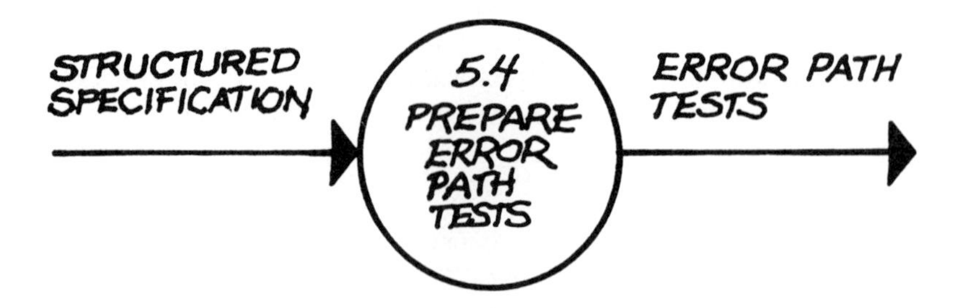

Figure 8.6 Prepare error path tests.

Where can one find such devious minds? Certainly not among members of the development group! The users may be a good source of error test cases, especially if they have had previous experience with the development of an EDP system. Unfortunately, users are often as unimaginative as developers when it comes to inventing nasty test cases. Unless your user begins cackling like Count Dracula as he specifies his test cases, you should probably assume that his test cases are too tame for you. University students are sometimes a good source of nasty test cases. Test data generators—computerized packages that generate test cases—can help by creating random test cases that no reasonable person would have devised.

In my experience, however, the most thorough error testing has generally not been provided by users, university students, or even automated testing packages. Rather, it has come from veteran programmers and systems analysts who are not connected with the project and whose job performance is directly related to the number of errors they find in the software they are testing. These are typically people who have been in the business long enough to know instinctively where to find the weak spots in any software simply because such weak spots have been found in their own software over the years.

8.2.5 Subactivity 5.5: Package Tests

The job of packaging tests is mostly clerical. As you can see in Figure 8.7, subactivity 5.5 oversees the collection of performance tests, normal path tests, the test plan, and error path tests to produce a quality assurance test set, which is the union of these four inputs.

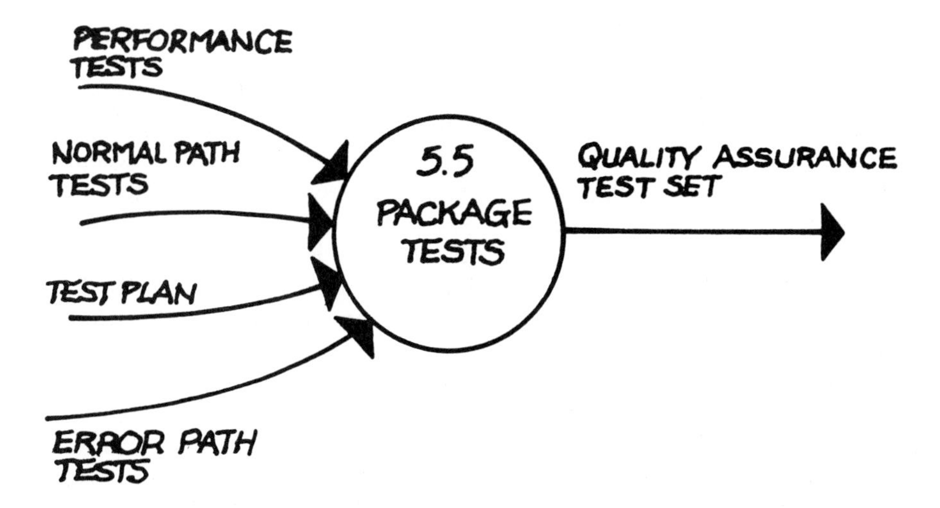

Figure 8.7 Package tests.

With the packaging effort, the acceptance test generation activity is completed. In the remaining pages of this chapter, let us look at the options and problems connected with the activity of acceptance test generation.

8.3 TOP-DOWN OPTIONS FOR ACCEPTANCE TESTING

In earlier chapters we have observed that the project manager has an almost infinite number of choices: He could elect a radical approach, a conservative approach, or something in between.

However, when it comes to acceptance testing, the options disappear. The acceptance test plan should coincide with the nature of the top-down approach followed in the other activities. Thus the acceptance test generation activity will be repeated again and again if the previous activities were performed in a radical fashion, or it may be performed once if the systems analysis, design, and implementation were performed in a conservative fashion.

8.4 PROBLEMS WITH GENERATION OF ACCEPTANCE TESTS

Almost anyone who has lived through a systems development project feels uncomfortable about testing. In retrospect, it always seems either that one did too much of it or, more likely, far too little of it. It seems either to go poorly or to go too well, with no errors detected until the system is delivered to the user! And it often seems that no matter how thorough the acceptance testing has been, the user still finds some reason for not liking the delivered system.

The problem of the user's not liking the system even if it has been properly tested is a problem that has preoccupied the EDP indusry almost obsessively for the past several years. After all, what is the point of exhaustive testing if we don't even know what the user wants us to be testing for because we don't really know what system the user wants? Proper systems analysis—using structured analysis and the activities described in Chapter 5 of this book—should help alleviate these problems. Unfortunately, however, even with the best systems analysis, there are still some common difficulties in the area of testing.

First is the problem of *not enough time* being devoted to the task of developing acceptance tests. In the best of all worlds, proper time and resources are allocated, and such allocation begins as soon as the structured specification is produced. However, if the user is not available to participate in the job of creating test cases and if the project manager (and the rest of the organization) is unaware of the importance of testing, acceptance test generation almost certainly will be done by members of the project team *after* the design, coding, and top-down integration are almost complete. By then the project may be behind schedule, the deadline for installation may loom near, and the project manager may feel enormous pressure to finish the job. In such an environment, it may well be the user who generates the test cases—after the system is in operation!

Next is the problem that *test cases may be created by the development team.* This is an obvious and common danger, especially with the scenario suggested in the preceding paragraph. Unfortunately, the developers are likely to incorporate the same errors and omissions into their test cases that they incorporated in their code.

It also happens that the *criteria* for successful testing *may not be properly defined.* In many cases "successful" testing simply means that everyone ran out of time, that everyone was sufficiently impressed with the amount of computer time that was consumed in the testing process, or that everyone was convinced that having already found *N* bugs, there couldn't possibly be any more.[4] In fact, none of these occurrences represents a satisfactory measure of success; our discussion of subactivity 5.1 suggests some of the criteria that might be used.

Finally, it may be *difficult to generate acceptance tests.* If the system obtains its inputs from a batch device using tape, cards, disk, and so on, it is a simple matter to prepare test inputs. However, many systems today expect their input from an on-line, interactive environment—for example, from a human typist at a time-sharing terminal. Unless the vendor's software allows for simulated terminal input from a batch device, proper acceptance testing is almost impossible. If the acceptance tests must be manually typed from a terminal, the volume of testing will almost certainly be insufficient. More important, it will be virtually impossible to use the same sequence of test cases (without typos and other human-introduced changes) over and over again, a repetition that is absolutely essential for proper debugging. If the system is going to be developed on a large mainframe CPU, with a vendor-supplied telecommunications monitor, the testing group can be reasonably certain that simulated on-line input (e.g., from a vendor's remote terminal emulation program) will be feasible; however, if the system is going to be developed on small minicomputers or a network of PCs, such simulated input may not be feasible.

[4]In fact, studies of real-world testing experiences indicate that just the opposite is true: The more bugs one finds in a computer program, the more (as yet undiscovered) bugs there are likely to be. As Tom DeMarco points out in *Controlling Software Projects* (1982), "When you see a cockroach in a restaurant, you don't say. 'There goes *the* cockroach.' Instead, you say, 'This place is infested!' "

8.5 SUMMARY

I think that it is entirely appropriate that the most popular book on software testing in the 1980s should be titled *The Art of Software Testing* (Myers, 1979). It *is* an art—and for most of us a black art. I think that the best you can do, as a manager, is adopt a defense posture for dealing with this black art. That means doing the following:

- Invest as much time and effort as you can in thorough testing of your new system.
- Preclude as many errors as possible by investing more time and energy in proper systems analysis, design, and coding techniques. Barry Boehm of TRW has pointed out in a classic study that it costs ten times less money to detect and corrrect a systems analysis error in the systems analysis phase of the project than to let the same error remain undetected until the design phase.[5]
- Use automated mechanisms—for example, test data generators, as well as "analyst tool kit" workstations—as much as possible to increase the efficiency of the testing process.
- Measure what your people are doing. Keep careful records of the number of test cases generated, the number of errors found, the density of errors within individual modules, and the number of errors attributed to individual programmers. Without such measurements, there is no hope of ever changing the business of testing from an art into a science.

Throughout this book, we have emphasized the notion of carrying out several activities of the project life cycle in parallel. The job of generating acceptance test data is perhaps the only activity that has always been recognized as one that could be overlapped with design, coding, and "local" testing. The acceptance test cases are often developed by users and systems analysts who were involved in the analysis but who have had no part in the design and coding—and since both groups work from the structured specification, it is fairly clear that they can work in parallel.

As we will see in the next chapter, the output of activity 5—a set of acceptance tests—combines with the integrated system produced by activity 4 and becomes input to the quality assurance activity. Quality assurance (if successful) produces an accepted system, which is merged with the outputs of database conversion and procedure description to form inputs for the final activity, installation.

[5]B. W. Boehm, "Software Engineering," *IEEE Transactions on Computers,* no. 12 (December 1976), 1226–1241, reprinted in Yourdon (1979).

9

Activities 6–9: The Final Activities

9.1 INTRODUCTION

In this chapter we consider the final four activities in the project life cycle: quality assurance, procedure description, database conversion, and installation. Each activity is important and may require a massive amount of work, but our discussion will be brief. Our main concern is to show how the structured techniques, the top-down iterative approach, and the other activities in the life cycle affect these final activities.

You wil note that subsections dealing with specific comments and problems associated with these final activities are included for activity 6 only. The remaining three activities consist of elements for which no additional characteristics need to be cited separately.

9.2 ACTIVITY 6: QUALITY ASSURANCE

In essence, activity 6, *quality assurance,* is simply the execution of the acceptance tests developed in activity 5. The context of quality assurance is shown in Figure 9.1; it has no subactivities. The desired output of activity 6 is an accepted system.

Some organizations define quality assurance much more broadly as an activity that takes place throughout the entire project. In this broader context, quality assurance could be interpreted to mean such things as these:

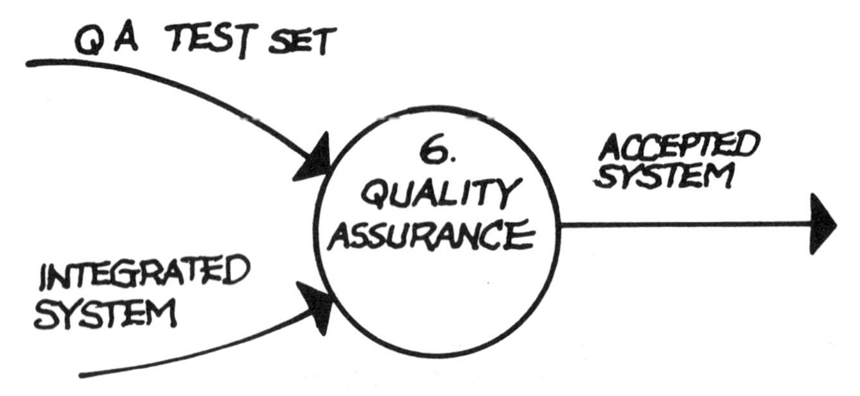

Figure 9.1 Context of quality assurance.

- Examination of the documentation associated with the project to ensure that it is complete, readable, and of an appropriately high standard
- Examination of the design and architecture of the system to ensure that it is adequate
- Examination of the source code to ensure that proper programming standards have been followed
- Examination of all aspects of the system from the viewpoint of security, auditability, and so on

There is certainly nothing wrong with this broader definition, and it does not contradict the approach taken in this book *as long as these broader activities are part of the specification of the system.* Thus we could imagine a statement of user requirements that says, "Please build a payroll system that carries out the following functions, and please build it to conform with the company's standards, as described in Report 367J."

9.2.1 Comments on Quality Assurance

I define quality assurance as the final opportunity for the user to voice his dissatisfaction with the system or for the developers to discover that they misunderstood the structured specification. In other words, it is the last chance for nasty surprises, which are precisely the sort of surprises you would desperately like to avoid! To help prevent nasty surprises, I suggest that you keep in mind that *all acceptance tests should be derived from the specifications,* that *acceptance tests should be binary,* and that *if quality assurance fails, it is probably the analyst's fault.* Let us discuss each point in turn.

All acceptance tests should be derived from the structured specification. Indeed, almost all of the criteria for acceptance that cannot be so derived should be explicitly stated elsewhere (typically in the physical constraints document in the structured specificaiton, for as DeMarco puts it so succinctly, "The specification *is* the acceptance test.").[1]

Keep in mind as well that the acceptance test should be binary. Either the system is acceptable or it isn't. Obviously, a few minor glitches are to be expected, but if the execution of the acceptance test cases produces a list of 7295 errors, the ball game is over. If you hear about projects in your organization in which acceptance testing has gone on for a year and a half, you know that "acceptance testing" has become a euphemism for "keep stalling until we can find some graceful way to bury this wretched project!"

Finally, if quality assurance fails, it's probably the analyst's fault. It's easy to blame the failure on the programmers—after all, by this stage they're the only ones who are left to accept the blame! In most conventional projects, the analysts have long since departed and are now working on Next Year's Glorious Management Information System, so it might not occur to management that the *real* blame for the failure of accpetance testing probably should go to the analysts.[2] Conversely, if the project is developed from the top down, there is a much greater chance that the analysts will still be around toward the end of the project: In the most radical top-down approach, coding *and* specifications are still being written on the last day of the project.

9.2.2 Problems with Quality Assurance

Most of the problems encountered during quality assurance have nothing to do with the conduct of quality assurance per se but rather with the fact that final testing uncovers problems of analysis, design, and implementation. However, two potential problems are associated with the actual conduct of the quality assurance activity; one deals with objectives definition, the other with documentation of errors.

As we discussed, the fundamental purpose of quality assurance is to use the test cases produced during acceptance test generation to ensure that the system performs according to the specification (no more, no less). But problems can occur because the objectives of the quality assurance activity may not be properly defined. Unfortunately, many people feel that the quality assurance activity is an open invitation to examine the system for possible deficiencies in areas that were never discussed in the specification. If, for example, they look for sufficient comments in the source code, proper design and implementation (from the viewpoint of the computer operations staff),

[1]DeMarco (1978), p. 326.

[2]This comment comes from the experience of many projects: Typically 50 percent of the errors associated with a systems development project, and 75 percent of the cost of error removal can be traced to errors of systems analysis.

of good structured coding techniques, inclusion of proper audit trails and audit controls, and so on, they must be certain that the criteria for these were included in the specification.

The fact that many systems do not have adequate comments, proper audit trails, or proper use of structured coding is not an excuse for a last-minute witch hunt to see if the system should be accepted. Operations standards, coding standards, documentation standards, and anything else that may eventually reflect on the system's acceptability and quality should be included in the structured specification. If it's not in the specification, it should not be an issue in quality assurance.[3]

A second potential problem is that the errors discovered during quality assurance are rarely documented for use in future projects. Activity 5 is where the test cases are created, but in activity 6 they are actually applied to the system. Thus it is in activity 6 that we learn how many errors there actually were within the system and of what type. It is also in activity 6 that we see how difficult it is to correct each error and how many *new* errors are introduced as a result of fixing an *old* error. All of this is the result of feedback (not shown in the project life cycle) from the quality assurance activity to the implementation activity. This information about the quantity of errors, the nature of the errors, and the difficulty of fixing them can be of tremendous use to project managers when they develop their plans, budgets, and schedules for future projects. Unfortunately, the information is rarely recorded anywhere, since it appears to have little use to the project at hand.[4]

9.3 ACTIVITY 7: PROCEDURE DESCRIPTION

Procedure description, the seventh activity in the system life cycle I've described, involves the translation of a specification (analyst's *and* designer's specifications) into a user manual. Its context is shown in Figure 9.2.

It is important to emphasize that *both* analysis and design products are required for the user manual, although most of the information will come from the analyst's specification, as that is the document that describes user policy and formally defines inputs and outputs. However, the analysis document is largely formatless, for the user typically has no policy regarding the detailed physical formats of CRT screens, report layouts, and other inputs and outputs. But although he may not care about such formats during analysis, he obviously must know what they are in order eventually to be able to use the

[3]As mentioned earlier, this can be accomplished by incorporating the standards (or a reference to a standard manual) in the specification itself.

[4]One solution to this problem is to form a "metrics team" whose job is to collect objective measurements about all aspects of a project. For more detail on this concept, see DeMarco (1982).

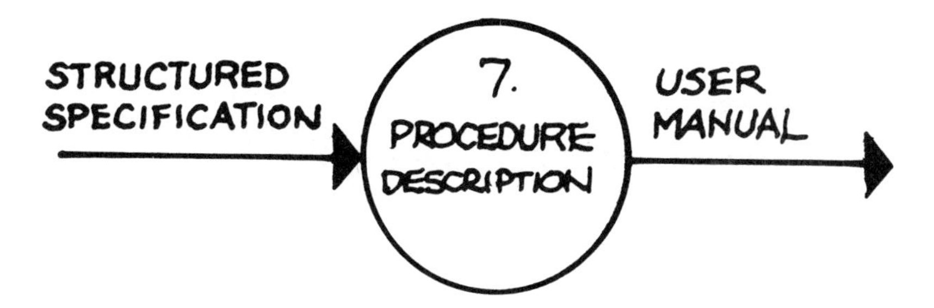

Figure 9.2 Context of procedure description.

new system. And since the physical formats are determined during the process of design, it follows that the design specification is an input to activity 7.

I have stated repeatedly in this book that the activities in the project life cycle need not be conducted in a linear, sequential fashion; this is particularly true of activity 7. Once the structured specification has been completed, it is possible for activity 3 (design), activity 5 (acceptance test generation), and activity 7 to commence simultaneously. As we have noted, activity 7 requires the output of activity 3 (the design specification), but it needs that output to *complete* its work, not necessarily to begin it. The bulk of the user manual can be written from the structured specification alone; detailed formats and other physical information can be filled in later.

Inded, we should carry this point about concurrent activities even further. We noted in Chapter 7 that the project team may build several skeleton versions of the system; that is, top-level modules may be coded and tested before low-level features of the system have been designed or even specified. One reason for doing this is to produce a demonstration for the user at the earliest possible time—a primitive version of the system that he may be able to use in an experimental fashion. But to make *any* use of the system, the user will require some kind of user manual; hence we might imagine a version 0 user manual that would accompany the version 0 code. This first user manual might be little more than an outline, with gaping holes to be filled in later when the details of analysis and design are complete. Development of the user manual would thus continue in a top-down fashion, in synchronization with development of the code for the new system.

9.4 ACTIVITY 8: DATABASE CONVERSION

The context of database conversion is shown in Figure 9.3; it has no sub-activities. In some projects, activity 8 may not exist. If, for example, the new system does not replace an existing system, it follows that there will be no existing database to convert. Alternatively, activity 8 can be a monumental

Figure 9.3 Context of database conversion.

task; for example, the user's current system may have a database of 10 million records that must be converted to a different format and perhaps an entirely different kind of file organization.

Although database conversion may be a large task, it is typically characterized by a need for massive amounts of CPU time rather than great intellectual capacity. That is, the conversion process is typically an automated one, and the major problem is often that of negotiating with the operations manager for sufficient computer time actually to perform the conversion!

In some cases, however, the user's existing database is not computerized; it consists of index cards, carbon copies of invoices, or file folders stuffed with bits and pieces of paper. When this is the case, there is the question of whether the existing database should be converted all at once or incrementally, transaction by transaction, once the new sysem is operational. The project life cycle presented in this book strongly implies the former, but both options should be considered by the project manager and the user.

If the existing database is manual, there is a good chance that at least some of it is missing, illegible, inconsistent, or ambiguous. Converting it to a new database will require a human being to read the existing data; make some attempt to understand, correct, and interpret it; and then regurgitate it in some form acceptable to the new system. For the sake of consistency and accuracy, it is usually best to have this done by a single person or a single group of people.

A complete conversion of the existing datdabase may require such a massive expenditure of time, energy, and money that the user will be reluctant to invest the resources to accomplish it. Note that the people who actually do the work will probably have to come from the user organization, since interpretation and understanding of large amounts of idiosyncratic data will be involved. The user will be even more reluctant to convert the database en masse if he senses that much of it is inactive—for example, if his database consists of 100,000 customers, but only 1000 are active in any given month.

Thus there may be a strong motivation to make activity 8 part of the actual operation of the system. The user must make the final choice here, but he should be aware of two disadvantages of the incremental approach. First, the incremental approach is more likely to lead to errors of style and substance in the conversion process, for the conversion will be carried out by several disparate users over a long period of time. Second, it will make the clerical-level users of the system appear *less* productive for the initial period of operational use of the system, that is, until the entire database has been converted. This is likely to be extremely annoying not only to those users but also to their immediate supervisors—the ones who were told, rather glibly, that the new system would be much more efficient and productive than the old one!

9.5 ACTIVITY 9: INSTALLATION

The context of activity 9 is shown in Figure 9.4. Installation is the end of the road, it is the final joining together of the software, user manual, and new database. In a small project, installation should be almost anticlimactic. If the previous activities have been carried out properly, the project should end not with a bang but with a whisper. The system is officially declared operational, the users begin typing transactions on their terminals, and everyone lives happily ever after.

Alas, it doesn't usually happen this way on large projects. Installation occurs when the project team finally emerges from the back rooms of the EDP organization and enters the world of reality. For the user, it's the moment when he realizes that he is no longer dealing with a skeleton system, to be dabbled with during his lunch hour; now it's a real system, and its performance will very likely have a significant impact on his day-to-day operations.

In other words, installation is often frightening. The prospect of cutting over to a system that may deal with thousands of on-line users in hundreds of different locations, a system that may have to run 24 hours a day, processing millions of transactions a year, is enough to make even veteran project

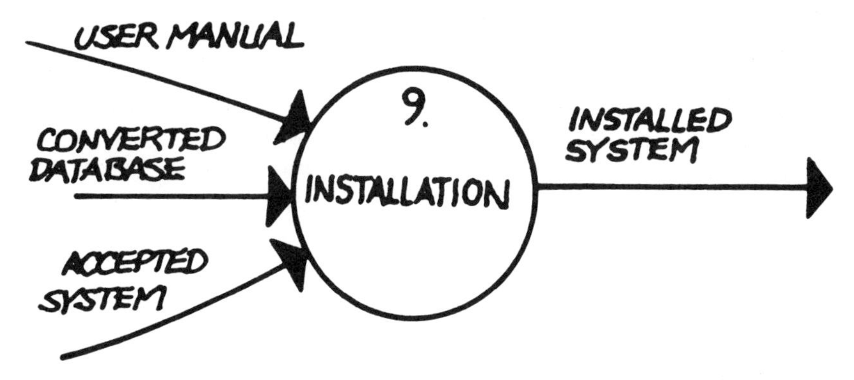

Figure 9.4 Context of installation.

managers a little nervous. And it is certainly enough to make the user more than a little nervous.

The details of implementation are unique for each project; however, the following issues will almost certainly have to be addressed. First, the detail of *when* the system will be installed must be settled. Careful planning is usually required to ensure that the installation of the new system does not disrupt the user's day-to-day business. Thus installation may have to take place in the middle of the night, at the opening of business in the morning, at the beginning of a fiscal year, and so on.

Second, one must determine when the old system will be dismantled. No matter how thorough the testing in activity 5, the user will almost certainly insist that the new system be run in parallel with his old system for some period of time. If he didn't, the programmers and analysts would probably have heart attacks—the old system is their safety net, in case something goes wrong! The user also realizes that the parallel operation is frustrating (especially for the person who has to verify that the new system is operating properly, even though the logical functions and physical formats of the new system are problably different from the old system) and, most of all, expensive! But the user knows that pulling the plug on the old system is probably an irrevocable act: Hardware will be dismantled, his old database will immediately become obsolete, and his people will almost immediately forget how they operated the old system.

A third question to be answered is whether the system should be installed all at once or in pieces. Consider a national bank in a country like England, Canada, or Australia: It has literally *thousands* of branches in half a dozen time zones. If it builds a new EDP system, should it attempt to install the system simultaneously in all its branches? The very thought is enough to boggle the mind! In such a case, installation will almost certainly have to proceed in stages over a period of one to two years—in contrast to a small system that can be installed in one short period of time.

Finally, it is essential to decide how the users should be trained. On a small project, training can be informal and ad hoc. The system is usually intrinsically simple, the user manual is adequate, and the development team is available if questions or problems arise. On a large project, such an approach would obviously be inappropriate. Indeed, the job of training could be so large and complex that it might form an entire activity in the project life cycle, with a modified version of the production software and a special version of the production database on which the users could practice.

9.6 SUMMARY

The fact that these last four activities have been discussed rather briefly does not diminish their importance. Quality assurance, procedure description, database conversion, and installation are all major and important activities, but the way these tasks are done has not changed as much as the way that analysis, design, and implementation activities have changed in the past decade.

It is important, of course, constantly to keep in mind the idea that several activities can take place in parallel. For example, activity 7 and activity 8 (procedure description and database conversion) can begin when activity 2 (analysis) has finished; similarly, some of the early planning for installation can begin as soon as there is a good idea of the user's requirements.

The final activity, installation, is the last one that I have chosen to discuss in relation to managing the system life cycle. I have not addressed the issue of operating the new system, maintaining it, or enhancing it. This omission is deliberate, for much of the work of maintenance and system enhancement is of the sort discussed throughout the book—and thus amenable to the project life cycle that applies to development projects.

Now that we have discussed the primary activities of a development project, from survey through installation, it is appropriate to take one last look at the big picture of the project life cycle. This last overview, including some final comments on the application of structured techniques to EDP projects, is the subject of Chapter 10.

10

Final Observations

We conclude our presentation of the structured project life cycle with a discussion of several general issues:

- Does the structured project life cycle solve all of the problems of project management?
- What percentage of the project time and resources should be devoted to each of the major activities?
- How can we ensure the quality of each activity in the project life cycle?
- How do we organize our work force to carry out the activities?
- What are the major themes of the structured project life cycle?
- How can the new project life cycle be introduced in a typical organization?

10.1 DOES THE STRUCTURED APPROACH SOLVE ALL THE PROBLEMS?

Obviously, the structured project life cycle presented in this book does not solve all management problems. Chapter 1 pointed out that a variety of classical project problems will not necessarily be solved with the structured project life cycle: recalcitrant users, impossible deadlines, incompetent programmers, and so on.

Similarly, this book has not given you any "magical" techniques for deriving accurate estimates for the next project you work on. I firmly believe that the only way you will ever be able to perform accurate estimates is to compile a substantial database of management statistics about *previous* systems development projects in your organization so that your estimates of manpower, budget, and time requirements for a new project can be derived from actual figures kept for previous projects. However, it is encouraging to note that more and more "general" models and estimating formulas are being developed in the industry. For more information, consult the work of DeMarco (1982), Jones (1986), and Boehm (1982).

Although the structured life cycle doesn't solve all political problems, it does—in my experience—dramatically improve the management process. First of all, some problems that appear to be political may turn out to be technical problems after all. Perhaps the reason your user community appears uncommunicative and uncooperative is that its members are simply unable to comprehend the classical functional specifications that your systems analysts have been giving them for the past decade; structured analysis should help. And perhaps the reason some programmers have appeared incompetent is that nobody ever showed them any rational methods for organizing and implementing their software; structured design and structured programming may help. And while the politics of impossible deadlines will always be with us, the concept of top-down implementation may help reduce the deadline battles between users, management, and the project team.

The structured project life cycle helps the management process in two other ways. It helps *define* the tasks that need to be carried out in an EDP project, and it provides *guidelines* and *methods* for carrying out those activities. Any project life cycle, for example, will identify the major activity of systems analysis; however, many fail to identify the subactivities. And even if the systems analyst is told that one of his tasks is to develop an essential model of the user's enterprise, he runs the risk of failure unless he has tools and guidelines for carrying out the task.

10.2 HOW MUCH TIME SHOULD BE DEVOTED TO EACH ACTIVITY?

Obviously, the amount of time spent on systems analysis, design, and implementation will vary from one project to another. On some projects the most difficult task may be that of capturing the user requirements; subsequent design and implementation may be trivial. On other projects the systems analysis may be straightforward, but the technical issues of design and implementation (of, say, a process control system) may be overwhelming. Many business-oriented projects today find that the programming and implementaion phase of the project has been dramatically simplified by the introduction of fourth-generation programming languages. Thus it is dangerous even to suggest that there is a universal guideline for determining the amount of project resources that should be devoted to each activity.

In addition, we currently lack a database containing a large sample of systems development projects that have used the structured project life cycle. Numerous projects have used structured analysis *or* structured design *or* structured progamming *or* top-down implementation. However, projects that have followed the life cycle presented in this book are relatively few, and only a very small number of them have reported any statistics.

Nevertheless, from the statistics that I have been able to gather from individual projects and from experience with classical projects, I suggest the following division of labor among the various activities of the structured project life cycle:

Survey	5%
Analysis	35%
Design	20%
Implementation	15%
Remaining activities	25%

If these activities are carried out in the classical, sequential fashion, 60 percent of the project resources will be consumed before the first line of code is written. You must decide for yourself, of course, whether this is politically tolerable in your organization!

10.3 HOW CAN WE ENSURE THE QUALITY OF THE ACTIVITIES?

One of the biggest problems with any new technological development is that it may be perceived as a panacea—a form of magic that will allow people whose brains are in the off position to solve problems that had been heretofore unsolvable. A similar (and perhaps related) problem is that of people paying lip service to a new technology that has acquired the reputation of a panacea.

If you've been in the systems development business for more than a few years, you've probably learned to be suspicious of any panacea—and that includes, of course, the structured techniques. However, your technicians, particularly the younger and more impressionable ones, may not have learned the cautious skepticism with which you approach any new technological development. You have an obligation to make sure that your programmers and systems analysts (and your users and the managers above you) do not look on the project life cycle presented in this book as some new form of religion.

Equally important, you have an obligation to ensure that the techniques presented in this book are actually being followed. It's easy for a systems analyst to say that he is using structured analysis, but what he's actually doing may be something altogether different.[1] It's even easier for a programmer-analyst to say that he's using structured design or structured programming without his users or managers knowing whether it is an accurate statement because end users and nontechnical managers are not in the habit of looking at such technical products as a COBOL program or a structure chart.

One way of preventing this problem is to insist on formal walkthroughs and reviews at the end of each major activity of the project life cycle, at the end of each subactivity, and at the end of each level of decomposition within each subactivity. Thus the manager should insist on a formal review at the end of the systems analysis activity; within that, he should insist on a formal review at the end of the behavioral modeling subactivity. Indeed, he should insist on a formal review of the top level of the behavioral model, the second level, the third level, and so on. Several papers and books provide guidelines for conducting such formal reviews, and the bibliography at the end of the book provides several references. As important as the *methods* for conducting the walkthroughs, though, are the *participants*. The walkthroughs for your first structured project should include outsiders as well as appropriate members of the project team. You may wish to include members of other project teams, representatives of your organization's internal audit department, representatives of the organization's external auditing firm,[2] or representatives of other outside consulting firms. Whoever is chosen, make sure that the outside reviewers can give you an unbiased, objective opinion on the proper use of the techniques and procedures of the structured project life cycle. Also keep in mind that it is always a good idea to seek active user participation throughout the walkthrough and review process.

You should also consider the use of workstations and other automated tools. As I have pointed out elsewhere in this book, automated tools have the advantage of providing mechanical error-checking facilities. This will help, for example, to ensure that the data flow diagrams and structure charts produced by the systems analysts and designers are complete and consistent.

[1]This commonly happens on the *first* attempt to use structured analysis: The systems analyst often becomes very involved in the drawing of data flow diagrams but spends little or no time developing the data dictionary or the process specifications.

[2]Many of the "Big 8" accounting firms, as well as some of the smaller ones in the United States, Canada, and Europe, now have a fair amount of experience in the use of the structured techniques and may thus be qualified to help you in this area.

10.4 HOW DO WE ORGANIZE OUR WORK FORCE TO CARRY OUT THE ACTIVITIES?

The activities described in the structured project life cycle will, for the most part, appear familiar to most project managers. Systems analysis is done by systems analysts; design, by designers; programming, by programmers; and so forth. Size of the project, and size of the staff, may lead to separate groups doing each of these activities, or it may lead to a single group "changing hats" from systems analysis to design to programming.

However, one aspect of the structured project life cycle has had an enormous impact on the organization of the work force for EDP projects: the emphasis on high-level tasks and low-level tasks within *each* of the activities of the structured project life cycle. Thus the systems analysis phase has high-level activities that require the talents of someone with ten years or more of experience; it also includes low-level activities (for example, the writing of process specifications for some low-level portions of a system) that could be accomplished by junior technicians having less than a year of experience.

In classical projects, the manager generally assigns his most senior staff to the task of systems analysis, his less exprienced staff to the task of systems design, and his most junior people to the task of programming. There are two problems if this approach is used within the structured project life cycle: The senior people will noisily object to the amount of detailed, tedious work they are being asked to do, and the junior people will complain loudly that virtually all of the creativity has been taken out of their work.

The solution, as many project managers have found, is to assign senior personnel and junior personnel to *each* of the major activities in the structured project life cycle. Senior personnel can interview users and draw the data flow diagrams in the analysis activity; junior personnel can concentrate on the details of the data dictionary and the process specifications. The junior person who writes a process specification for a small piece of the user's business requirements can also be responsible for translating that process specification into a design specification (written, perhaps, in pseudocode) and then finally into COBOL, FORTRAN, C, Pascal, or Ada in the implementation activity. This may cause some political problems within the organization, for it implies that senior technical people with 15 years of experience may write the code for top-level modules in a system when they thought they had escaped the job of coding forever, but I believe that it is a very healthy development.

10.5 WHAT ARE THE MAJOR THEMES OF THE STRUCTURED PROJECT LIFE CYCLE?

The major themes of the structured project life cycle have been emphasized throughout this book. But they are important, so let me summarize them here:

- *Modeling.* Inexpensive graphic models of the requirements, the design, and the code allow us to investigate the properties of a system before we invest large sums of money. This book has concentrated on *paper* models; however, other types of models (prototypes, simulations, etc.) may be equally appropriate in some cases.
- *Partitioning.* Virtually everything about a modern information system is too complex for a single person to grasp. The code, the design, and the underlying business requirements all benefit from scrutiny that is facilitated if the system is partitioned in such a way that one can first obtain an overview and then progress in an orderly top-down fashion all the way down to the microscopic details.
- *Iteration.* Nothing complex is ever done perfectly the first time. Our methods of systems analysis, systems design, and implementation must allow for a mediocre beginning and several opportunities to make incremental improvements. However, it is also important for the project manager to realize that ongoing iterations will eventually reach a point of diminishing returns; one must eventually stop "gilding the lily" and move onward to the next activity.
- *Parallel activities.* A project life cycle that insists that all of the available resources be devoted to one activity at a time will not work in the real world. Such an approach works only on projects in which the surrounding environment is not only stable but also virtually static. We *must* allow some overlapping of systems analysis, design, implementation, test data generation, and all the other activities in the life cycle. The degree of overlap should be determined on a dynamic basis as a matter of negotiation among the project manager, the user, the project team, and the higher levels of project management.

10.6 HOW SHOULD THE LIFE CYCLE BE INTRODUCED?

Now that the methods and techniques for managing the system life cycle have been presented, I would like to use these last paragraphs to address the most important question of all: Under what circumstances should the structured project life cycle be introduced in your organization? Should it be adopted by all project managers, throughout the entire organization, on the same day? Should it be imposed, by decree, by the top-level EDP manager?

The problems of introducing a new project life cycle are in many ways similar to the problems of introducing the individual components of the life cycle (structured analysis, structured design, etc.). The implementation strategy for individual structured techniques and for the introduction of an overall structured project life cycle may be summarized as follows:

- Don't impose a new project cycle by decree. Many project leaders and technicians will resist if they feel that they have no voice in the matter.
- Avoid the approach of training everyone in the organization in a brief half-day seminar and then letting them flounder on their own. This kind of superficial training usually causes more harm than good. Thorough training, supported by appropriate consulting assistance, is important.
- Finally, try the new approach on a pilot project—or several pilot projects, if appropriate—and use the people who participate in the pilot to provide real-world guidance to subsequent groups who adopt the new techniques. However, don't "bet the company" on your pilot project; choose a relatively small, safe project as your first experiment.

BIBLIOGRAPHY

BAKER, F. T., "Chief Programmer Team Management of Programming," *IBM Systems Journal,* 11, no. 1 (January 1972) 56–73. (Reprinted in Yourdon, 1979, pp. 65–82.)

BLOCK, R., *The Politics of Projects.* New York: Yourdon Press, 1983.

BOAR, B., *Application Prototyping.* New York: Wiley, 1984.

BOEHM, B., *Software Engineering Economics.* Englewood Cliffs, N.J.: Prentice-Hall, 1982.

BROOKS, F. P., *The Mythical Man-Month.* Reading, Mass.: Addison-Wesley, 1975.

COUGER, J. D., AND R. W. KNAPP, eds., *Systems Analysis Technqiues.* New York: Wiley, 1974.

COUGER, J. D., M. COLTER, AND R. W. KNAPP, eds., *Advanced Systems Analysis/ Feasibility Techniques.* New York: Wiley, 1982.

DAHL, O. J., E. W. DIJKSTRA, AND C. A. R. HOARE, *Structured Programming.* Orlando, Fla.: Academic Press, 1972.

DEMARCO, T., *Structured Analysis and System Specification.* New York: Yourdon Press, 1978.

———, *Concise Notes on Software Engineering.* New York: Yourdon Press, 1979.

———, *Controlling Software Projects.* New York: Yourdon Press, 1982.

DICKINSON, B., *Developing Structured Systems.* New York: Yourdon Press, 1981.

FAGAN, M. E., "Design and Code Inspections to Reduce Errors in Program Development," *IBM Systems Journal,* 15, no. 3 (July 1976) 182–211. (Reprinted in Yourdon, 1982, pp. 123–148.)

FLAVIN, M., *Fundamental Concepts of Information Modelling.* New York: Yourdon Press, 1981.

GANE, C., AND T. SARSON, *Structured Systems Analysis: Tools and Techniques.* Englewood Cliffs, N.J.: Prentice-Hall, 1979.

GILDERSLEEVE, T., *Successful Data Processing Systems Analysis.* Englewood Cliffs, N.J.: Prentice-Hall, 1978.

IEEE Transactions on Software Engineering, SE-3, no. 1 (January 1977). Entire issue devoted to structured analysis.

JACKSON, M. A., *Principles of Program Design.* Orlando, Fla.: Academic Press, 1975.

JONES, T. C., *Programming Productivity.* New York: McGraw-Hill, 1986.

KELLER, R., *The Practice of Structured Analysis.* New York: Yourdon Press, 1983.

KERNIGHAN, B. W., AND P. J. PLAUGER, *Software Tools.* Reading, Mass.: Addison-Wesley, 1976.

KING, D., *Current Practices in Software Engineering.* New York: Yourdon Press, 1984.

LISTER, T. R., AND E. YOURDON, *Learning to Program in Structured COBOL, Part 2.* New York: Yourdon Press, 1978.

MARTIN, J., *An Information Systems Manifesto.* Englewood Cliffs, N.J.: Prentice-Hall, 1984.

MARTIN, J., *Application Development Without Programmers.* Englewood Cliffs, N.J.: Prentice-Hall, 1984.

MCMENAMIN, S., AND J. PALMER, *Essential Systems Analysis.* New York: Yourdon Press, 1984.

MYERS, G. J., *Reliable Software Through Composite Design.* New York: Petrocelli/Charter, 1975.

———, *The Art of Software Testing.* New York: Wiley Interscience, 1979.

ORR, K. T., *Structured Systems Development.* New York: Yourdon Press, 1977.

———, *Structured Requirements Definition.* Topeka, Kans.: Kenn Orr & Associates, 1981.

PAGE-JONES, M., *The Practical Guide to Structured Systems Design.* New York: Yourdon Press, 1980.

———, *Practical Project Management.* New York: Dorset House, 1985.

PETERS, L. J., *Software Design: Methods and Techniques.* New York: Yourdon Press, 1981.

WARD, P., *Systems Development Without Pain: A User's Guide to Modeling Organizational Patterns.* New York: Yourdon Press, 1984.

———, AND S. J. MELLOR, *Structured Development for Real-Time Systems.* New York: Yourdon Press, 1985.

WARNIER, J. D., *The Logical Construction of Programs.* New York: Van Nostrand Reinhold, 1976.

———, *The Logical Construction of Systems.* New York: Van Nostrand Reinhold. 1981.

WEINBERG, G. M., *The Psychology of Computer Programming.* New York: Van Nostrand Reinhold, 1971.

WEINBERG, V., *Structured Analysis.* New York: Yourdon Press, 1978.

WEINBERG, G. M. AND FREEDMAN, D., *Handbook of Technical Inspections.* New York: Little Brown, 1977.

YOURDON, E., "A Case Study in Structured Programming: Redesign of a Payroll System," *Proceedings of the IEEE Compcon Conference, 1975.* New York: Institute of Electrical and Electronic Engineers, 1975.

———, *Techniques of Program Structure and Design.* Englewood Cliffs, N.J.: Prentice-Hall, 1976.

———, ed., *Classics in Software Engineering.* New York: Yourdon Press, 1979.

———, ed., *Writings of the Revolution: Selected Readings on Software Engineering.* New York: Yourdon Press, 1982.

———, *Managing the Structured Techniques* (3rd ed.). New York: Yourdon Press, 1985.

———, *Structured Walkthroughs* (3rd ed.). New York: Yourdon Press, 1985.

———, *Nations at Risk: The Impact of the Computer Revolution.* New York: Yourdon Press, 1986.

———, AND L. L. CONSTANTINE, *Structured Design: Fundamentals of a Discipline of Computer Program and Systems Design.* Englewood Cliffs, N.J.: Prentice-Hall, 1979.

———, C. GANE, AND T. SARSON, *Learning to Program in Structured COBOL, Part 1,* (2nd ed.). New York: Yourdon Press, 1978.

Index

S

T

U

W

Y

TEAR OUT THIS PAGE TO ORDER THESE OTHER HIGH-QUALITY YOURDON PRESS COMPUTING SERIES TITLES

Quantity	Title/Author	ISBN	Price	Total $
_____	Building Controls Into Structured Systems; Brill	013-086059-X	$32.00	_____
_____	C Notes: Guide to C Programming; Zahn	013-109778-4	$16.95	_____
_____	Classics in Software Engineering; Yourdon	013-135179-6	$37.33	_____
_____	Concise Notes on Software Engineering; DeMarco	013-167073-3	$17.00	_____
_____	Controlling Software Projects; DeMarco	013-171711-1	$36.33	_____
_____	Creating Effective Sofware; King	013-189242-8	$33.00	_____
_____	Crunch Mode; Boddie	013-194960-8	$27.00	_____
_____	Current Practices in Software Development; King	013-195678-7	$33.33	_____
_____	Data Factory; Roeske	013-196759-2	$22.00	_____
_____	Developing Structured Systems; Dickinson	013-205147-8	$32.00	_____
_____	Design of On-Line Computer Systems; Yourdon	013-201301-0	$47.00	_____
_____	Essential Systems Analysis; McMenamin/Palmer	013-287905-0	$32.00	_____
_____	Expert System Technology; Keller	013-295577-6	$26.95	_____
_____	Concepts of Information Modeling; Flavin	013-335589-6	$26.67	_____
_____	Game Plan for System Development; Frantzen/McEvoy	013-346156-4	$30.00	_____
_____	Intuition to Implementation; MacDonald	013-502196-0	$22.00	_____
_____	Managing Structured Techniques; Yourdon	013-551037-6	$32.00	_____
_____	Managing the System Life Cycle 2/e; Yourdon	013-551045-7	$33.00	_____
_____	People & Project Management; Thomsett	013-655747-3	$21.33	_____
_____	Politics of Projects; Block	013-685553-9	$21.33	_____
_____	Practice of Structured Analysis; Keller	013-693987-2	$26.67	_____
_____	Program It Right; Benton/Weekes	013-729005-5	$21.33	_____
_____	Software Design: Methods & Techniques; Peters	013-821828-5	$32.00	_____
_____	Structured Analysis; Weinberg	013-854414-X	$39.33	_____
_____	Structured Analysis & System Specifications; DeMarco	013-854380-1	$42.67	_____
_____	Structured Approach to Building Programs: BASIC; Wells	013-854076-4	$21.33	_____
_____	Structured Approach to Building Programs: COBOL; Wells	013-854084-5	$21.33	_____
_____	Structured Approach to Building Programs: Pascal; Wells	013-851536-0	$21.33	_____
_____	Structured Design; Yourdon/Constantine	013-854471-9	$48.00	_____
_____	Structured Development Real-Time Systems, Combined; Ward/Mellor	013-854654-1	$70.33	_____
_____	Structured Development Real-Time Systems, Vol. I; Ward/Mellor	013-854787-4	$32.00	_____
_____	Structured Development Real-Time Systems, Vol. II; Ward/Mellor	013-854795-5	$32.00	_____
_____	Structured Development Real-Time Systems, Vol. III; Ward/Mellor	013-854803-X	$32.00	_____
_____	Structured Systems Development; Orr	013-855149-9	$32.00	_____
_____	Structured Walkthroughs 3/e; Yourdon	013-855248-7	$23.33	_____
_____	System Development Without Pain; Ward	013-881392-2	$32.00	_____
_____	Teams in Information System Development; Semprivivo	013-896721-0	$26.67	_____
_____	Techniques of EDP Project Management; Brill	013-900358-4	$32.00	_____
_____	Techniques of Program Structure & Design; Yourdon	013-901702-X	$42.67	_____
_____	Up and Running; Hanson	013-937558-9	$28.67	_____
_____	Using the Structured Techniques; Weaver	013-940263-2	$25.00	_____
_____	Writing of the Revolution; Yourdon	013-970708-5	$37.33	_____
_____	Practical Guide to Structured Systems 2/e; Page-Jones	013-690769-5	$35.00	_____

Total $ _____

- discount (if appropriate) _____

New Total $ _____

OVER PLEASE ➡

AND TAKE ADVANTAGE OF THESE SPECIAL OFFERS!

a.) When ordering 3 or 4 copies (of the same or different titles), take 10% off the total list price (excluding sales tax, where applicable).

b.) When ordering 5 to 20 copies (of the same or different titles), take 15% off the total list price (excluding sales tax, where applicable).

c.) To receive a greater discount when ordering 20 or more copies, call or write:

Special Sales Department
College Marketing
Prentice Hall
Englewood Cliffs, NJ 07632
201–592–2498

SAVE!
If payment accompanies order, plus your state's sales tax where applicable, Prentice Hall pays postage and handling charges. Same return privilege refund guaranteed. Please do not mail in cash.

☐ **PAYMENT ENCLOSED**—shipping and handling to be paid by publisher (please include your state's tax where applicable).

☐ **SEND BOOKS ON 15–DAY TRIAL BASIS** & bill me (with small charge for shipping and handling).

Name ______________________________

Address ______________________________

City ______________ State ______________ Zip ______

I prefer to charge my ☐ Visa ☐ MasterCard
Card Number ______________ Expiration Date ______________

Signature ______________________________

All prices listed are subject to change without notice.

Mail your order to: Prentice Hall, Book Distribution Center, Route 59 at Brook Hill Drive, West Nyack, NY 10995

Dept. 1 D–OFYP–FW(1)